D0350362

COPYCATS AND CONTRARIANS

COPYCATS & CONTRARIANS

Why We Follow Others ... and When We Don't

Michelle Baddeley

YALE UNIVERSITY PRESS
NEW HAVEN AND LONDON

For information about this and other Yale University Press publications, please contact:
U.S. Office: sales.press@yale.edu yalebooks.com
Europe Office: sales@yaleup.co.uk yalebooks.co.uk

Set in Adobe Caslon Pro by IDSUK (DataConnection) Ltd
Printed in Great Britain by TJ International, Padstow, Cornwall

Library of Congress Control Number: 2018940284

ISBN 978-0-300-22022-3 (hbk)

A catalogue record for this book is available from the British Library.

10 9 8 7 6 5 4 3 2 1

To my parents, with gratitude

Let us boldly contemn all imitation, though it comes to us graceful and fragrant as the morning; and foster all originality, though, at first, it be crabbed and ugly as our own pine knots.

Herman Melville, 'Hawthorne and His Mosses' (1850)

Worldly wisdom teaches that it is better for reputation to fail conventionally than to succeed unconventionally.

John Maynard Keynes, *The General Theory of Employment, Interest and Money* (1936)

Contents

Introduction

On 6 September 1997, the funeral of Diana, Princess of Wales attracted a crowd of over 3 million mourners in London, as well as a worldwide TV audience of almost 3 billion. The metres-deep carpets of bouquets, poems, teddy bears and other sentimental offerings accumulating outside Buckingham Palace and Diana's Kensington Palace home gave the twentieth century some of its most iconic images. Millions of strangers expressed extreme – if short-lived – grief about the death of a person they had never met. Why did so many individual mourners feel deeply enough to join with millions of others in expressing their collective sadness? They joined together as a grief-stricken herd, coordinated around the globe by international news media. This powerful mass hysteria seemed as unreasoning as it was uncontrollable. But was it?

Our herding is not always histrionic. Our tendency to imitate, follow others and group together can be reasonable strategies to improve our lives and evolutionary life chances. Herding is an instinct we share with other animals too. Behavioural ecologists have observed clever copying behaviour amongst many of our close (and not so close) animal relatives.

One example was uncovered by behavioural ecologists studying the behaviour of a small Australian marsupial called the quoll. Its survival was being threatened by the cane toad, introduced to Australia in the 1930s in a misguided attempt to manage the destruction of sugar cane plantations by cane beetles. To a quoll, these toads look as tasty as they are poisonous, and the quolls who scoffed them suffered fatal consequences at a speedy rate. Behavioural ecologists identified a clever solution by constructively harnessing quolls' instincts to imitate. Small groups of quolls were trained to be 'toad-smart' via a form of aversion therapy. They were fed toad sausages spiked with harmless but nausea-inducing chemicals, conditioning them to avoid the toads. Groups of these toad-smart quolls were then released back into the wild: they taught their own offspring what they'd learnt. Other quolls copied these constructive behaviours through a process of social learning. As each baby quoll learnt to avoid the hazardous toads, so the chances of the survival of the whole quoll species – and not just that of each individual quoll – were improved. The quolls were saved via minimal human interference, because ecologists were able to leverage quolls' natural imitative instincts.[1]

Diana's mourners and the toad-smart quolls illustrate that, as social animals, we clearly have strong instincts to copy and conform, a pattern of behaviour that has helped many species, including our own, to survive and prosper. But this is only half the story. Humans are not conformists always and everywhere. There are plenty of rebels and contrarians, some of whom have changed lives and history. Socrates was a famous example: he was sentenced by a jury to death by hemlock in 399 BC as punishment for refusing to worship the gods revered by his fellow Athenians, for appearing to side with the Spartans, and for embracing a role as self-appointed critic and gadfly of the Athenian state. But while Socrates ended his life as an outcast, our intellectual history was transformed by his contributions. Similarly, our modern lives would be unimaginable if history had not delivered a wide range of different characters prepared to take maverick risks: from Copernicus and Galileo through to Darwin, Crick and Watson. Via careful thought and deliberation, these and other mavericks and mavens have led us down new paths, unimaginable and contentious at the time. The consequences of the risks they took with their reputations and social standing were profound in terms of transforming the length and quality of our lives.

Herding and anti-herding defined

What exactly is herding? And what is its opposite? The literatures on copycats herding is vast (though there is less emphasis on contrarians) and span a wide range of subjects and contexts. With such a diversity of researchers studying herding, a universal definition is likely to be elusive. But there are three common threads that unify conceptions of herding that we can observe in ourselves and other copycats around us. First, and most obviously, herding involves imitation. Second, it is a group phenomenon: someone imitating just one other person is not herding; many people imitating one person – and many people imitating many people – is herding. Third, herding may sometimes be driven by unconscious motivations, as we shall see, but it is not random. Conscious and unconscious forces encourage us to choose to follow groups in systematic ways. Pulling all these threads together, we can define herding as a systematic choice to copy others in a group. It may benefit the self-interested individual, or it may bestow a collective advantage if individuals are joining with their fellows to support the interests of groups and/or species.

Understanding herding copycats will also help us to understand the essence of their opposites: the contrarians. Contrarians are 'anti-herders', where anti-herding can be defined as a choice *not* to copy others in a group.[2] Anti-herding is not as dissimilar from herding as we might at first imagine. Anti-herding is a group behaviour, and it is not random; but it is the mirror image of imitation because an anti-herding contrarian acts against, not in concert with, the group. Further, anti-herding shares two of the three features of herding outlined above, but with a few twists. Anti-herding is often a group phenomenon because it does not concern behaviour that is random or orthogonal to the group's behaviour. Contrarians are not hermits. They worry what others think, but they may deliberately decide to oppose the herd – sometimes by leading the group instead of following it. Like herding, anti-herding is systematic, not random, and perhaps it is *more* systematic if it is driven by deliberate, conscious choices. Either way, the actions of anti-herding contrarians and herding copycats can be complementary, in both good and bad ways.

Another key characteristic of herding is that it is social behaviour. We have evolved to be social animals, an evolutionary path that has instilled in us instincts to group together, reinforced by the social skills learnt during infancy and childhood. We trust and cooperate with others, even with strangers many miles away from us. We are often altruistic and philanthropic, even though our kindness to others reflects a complex mix of self-interest and generosity. It is a two-way interaction. When others are kind to us, we reciprocate. And when we reciprocate we build trust, and not only with our family, friends and communities. Most of our daily activities, including our economic activities such as work and shopping, would not be possible without trust and reciprocity. All these phenomena are linked to our more outward-looking and gregarious sides. Myriad experiments from psychology, neuroscience and behavioural economics have verified our strong social instincts, instincts that are shared widely – across countries, cultures and other animal species, including our close primate cousins.

What has this to do with copycats and contrarians? Copying, herding and imitating are another facet of our social nature, and our herding tendencies complement these other aspects of our sociality. Crowds of like-minded people will gather together, in a political protest for example, because they share a level of trust – in each other and in the cause or leader that they are supporting. The same people would be as reluctant to join a crowd of opponents they do not trust as they are enthusiastic to join a crowd of people they do trust. Marketers and advertisers know well that if we can be persuaded that certain celebrities are trustworthy, then we can be encouraged to follow them by buying the products they endorse. Local and communal activities – from cake sales to charity auctions – are examples of how we bring together our desires to join a group with our generous and reciprocating natures.

Why herd?

The behaviours exhibited by Diana's mourners, quolls and Socrates may appear, superficially, to be different. Scratch the surface, however, and we can see they do share some commonalities in what they tell us about how and why we imitate others – and when we don't, why we don't. Many herding researchers from across the social and behavioural sciences have focused on capturing the social influences underlying our propensities to herd, and these can be roughly divided into the categories of *informational influences* and *normative influences.*

Informational influences include all the ways in which we learn by gathering information from others around us. What others do, and whether they succeed, is important information we can use to our own advantage. We observe how others choose and decide and this helps us to choose and decide for ourselves. We may also be able to see how others' choices work out for them – so we can learn from their mistakes as well as their successes. The Garissa University College attack in April 2015 was a powerful, but harrowing, example of how copying driven by social learning can save lives. Four gunmen from the al-Shabaab jihadist militant group stormed the Kenyan college. They took students hostage, showing mercy only if a student could convince them that they were Muslim by reciting a key tract from the Qur'an. Those who could not cite the relevant tract were shot. One Christian student watched what was happening to those in the line in front of her and quickly learned to memorise the tract she needed to recite in order to persuade the hostage-takers that she was a Muslim. She saved her own life through social learning, by gathering information about others' choices and their consequences. From the perspective of her fellow Christians, this social information led her towards anti-herding, not herding; but she had learnt that copying most of the other Christians in the line ahead of her was not going to ensure her survival.

Normative influences encompass the norms and customs that define the groups and communities around us. Our responses to normative influences are often less conscious and deliberate than our responses to informational influences. We copy others because we feel a compulsion from others around us to conform – reflecting social norms, peer pressure and groupthink. The queue, for instance, is a famously sacred British institution. Most Brits would not dream of pushing into a long queue or joining a free-for-all stampede to the front, even when it might obviously be in their best interests to do so and harmful consequences are unlikely. London's *Evening Standard* reported an engaging example: 200 of Ed Sheeran's biggest fans, who had bought tickets in an online frenzy for one of his 2017 London concerts, calmly and entirely voluntarily formed an orderly queue outside the venue without instructions. Neither physical barriers nor policing were needed.[3]

Figure 1. Voluntary queuing at an Ed Sheeran concert, 2017.

Like Sheeran's adoring fans, and without consciously thinking about it too hard, we are aware that we will violate social norms and incite disapproval from strangers if we appear to be pushing in and prioritising our own wants at the expense of the many around us. Some of us will be happily waiting patiently in queues; others might be exerting effort in controlling aggressive instincts to push in. Either way, the queue represents a cooperative solution that minimises discomfort for the crowd.

The different types of normative influence are diffuse and harder (if not impossible) to quantify, but they are just as important as informational influences. Possibly they are *more* important because they are ingrained, automatic responses that we do not consciously notice in ourselves. They can also, perhaps counterintuitively, help to explain contrarian behaviours: people who behave in unconventional ways are sometimes simply adhering to unconventional norms, shared by a small fringe of marginalised groups with which they identify.

Consequences

Neither rebellion nor conformity is inherently good. Neither is inherently bad. If we follow others in buying into a rising housing market, for example, we may do very well out of gains in our property's value. If we follow others out of a collapsing football stadium, then we risk death by trampling. If we lead others out of a burning building, then we may all survive. If contrarians lead others into war, terrorism or gang violence, then they are risking others' lives, and sometimes their own. Even in terms of universal virtues, we would find it difficult to come to a clear conclusion. Copycats and contrarians are driven by the tensions between exploiting and using the group versus belonging and contributing to the group. Copying and herding manifest themselves in a wide range of ways: individuals operating in their own self-interest; collectives of individuals working together as a team towards a shared goal; madding crowds which seem to have a life and mind of their own and in which each individual person is dispensable. And even as individuals, we are not consistent. We all have the capacity to be copycats in some situations and contrarians in others. In our social and cultural lives, whether we decide to be copycats or contrarians will be determined by our different identities, formed by different contexts and our different roles in society. Like Dr Jekyll and Mr Hyde, a person who is conventional, diligent and professional during the daytime may be more unconventional, rebellious and disruptive at night.

Where do we begin in understanding all these complex interplays? The simplest place is by looking at what drives each of us as self-interested individuals to copy others and join groups. Economists have explored this theme extensively, focusing on how social learning and other rational motivations might encourage us to join a herd, as we shall see in chapter 1. From there we will fill in the many gaps in the simple economic model by looking across the social and biological sciences for other insights that can help to explain our copycat and contrarian natures.

What are the implications for our everyday lives? Some are worrying. Our evolved instincts, personalities, even our aptitude for intelligence can help to explain copycat and contrarian attitudes, choices and behaviour. But those evolutionary qualities are not necessarily a good fit in today's world. We live in an age in which we are digitally and globally interconnected in ways that could not have been imagined even a century ago, let alone when modern humans were evolving many hundreds of thousands of years in the past. Where does the group begin and end? When should we use information implicit in a group's actions and when should we ignore it? Ancient evolved animal behaviours operating within our artificial modern world can incubate a range of perverse behaviours, herding included. Our inclinations to copy or rebel do not always fit well with social media echo chambers, volatile stock markets, sensationalist clickbait newspaper reporting, political populism and information overload.

In the many volumes of papers and books about herding and contrarianism, writers and researchers tend to zero in on subject-specific research questions. This book is different. It brings together insights from a broad range of studies in a multidisciplinary account. Some economic theories explore why, as self-interested individuals, we might feel inclined to herd or rebel. From psychology and sociology we can see that unconscious social influences are powerful, but copying them does not always work out well. Neuroscience, evolutionary biology and behavioural ecology can give us some understanding of where our copycat and contrarian instincts come from, and how they play out in our everyday lives. All of these insights together can answer some pressing questions. What are the origins of our copycat and contrarian instincts and inclinations? How do copycats and contrarians interact? Do our copycat and contrarian instincts equip us well in the modern world? And, perhaps the most important question of all: what can we do about it?

1

Clever copying

Are copycats clever? Or is it mindless and irrational just to do what others are doing without using our own initiative and mental energy to decide for ourselves? And how might we distinguish blind conformity from intelligent imitation? Often, we cannot easily tell the difference.

Our everyday lives provide some examples. Imagine that you are in a meeting and you are asked to vote on an issue about which you do not feel particularly passionate or well informed. You decide to raise your hand in favour because you see a few of your colleagues doing the same. Are you being lazy? Responding to peer pressure? Perhaps. Or perhaps you are using your colleagues' actions as an alternative source of information. You interpret their hand-raising as a signal that they know something you do not. If you knew what they know, then perhaps you would vote in favour too. In cases like this, following others is clearly not stupid, even when it involves minimal brainwork.

At times, all of us find it easier just to follow what others are doing because we assume they know more than we do. When we are lost, it is reasonable to follow a crowd to find our way. By observing and following the actions of others, we can gather signals, information and guidance – all of which can help us to do better for ourselves. This is the phenomenon of self-interested herding. We herd because we get some benefit as selfish individuals.

Rational choice theory, developed in the 1970s by the Nobel Prize-winning American economist Gary Becker, provides deeper exploration of what motivates individuals to follow a crowd or join a group.[1] Becker maintained that individuals are the best at choosing for themselves. No other person or organisation is better able to prioritise the individual's interests in a rationally analytical way. Becker and his colleagues argued that this assumption helps to explain a wide range of human decisions and problems – everything from marriage and divorce to addiction and discrimination. Becker's rational choice approach is most commonly embraced by economists, with many economic models describing rational individuals making choices to help themselves, as if guided by sophisticated mathematical rules. Even so, Becker does allow that social interactions are important to us. He argues that our social environment has monetary value, helping us to generate what he calls 'social income' via our relationships with others around us.[2] Our professional relationships illustrate Becker's point: the opinions of our colleagues and bosses may have monetary value for us if they increase our chances of a pay rise.

To explain self-interested herding, economists start with the concrete advantages each person might enjoy from following others. A self-interested person is not concerned with promoting group interests. From an economist's perspective, we do not herd to help the group; we herd to help ourselves. We can learn from others. Sometimes, we can improve our reputations by following others. We can gain more when we act as a group than as an individual. All these advantages can be understood relatively easily in terms of economic motivations and incentives. Copying and collaborating is a means to an end – the end being something to do with helping ourselves.

Homo economicus in the crowd

How do economists link their assumptions about our capacity for rational choice with human social behaviour? Some insights have their roots in the ideas of Vilfredo Pareto, an Italian polymath who trained as an engineer and went on to make a range of enduring contributions to economics, sociology and political science. Often lauded as one of the forefathers of modern neoclassical economics, Pareto was one of many characters contributing to an impressive Italian tradition in economic analysis which has included inspirational thinkers from both left and right.[3] Pareto's name is well known to students of economics, associated as it is with one of the fundamental concepts in the subject – Pareto optimality. This is achieved when welfare improvements from voluntary exchange are exhausted, at the point where no-one can be made better off without making someone else worse off.

To ensure this simple (some would say simplistic) result, Pareto assumed rational choice by a peculiar, hypothetical species: *Homo economicus*.[4] Characterised by its clever, self-interested and individualistic nature, the choices *Homo economicus* makes are driven by rigorous, analytical decision-making processes as it searches for ways to maximise its own welfare. *Homo economicus* is not, however, infallible. It makes mistakes, but these are quickly corrected to ensure that they are not repeated. *Homo economicus* does not care what happens to others, but it does have a restricted social awareness. It realises that the information others convey in their choices and decisions is potentially useful, and it uses this social information to guide its choices, without necessarily worrying too much about how its actions may impinge on others' well-being.

What are the impacts on the economy as a whole? They are beneficial, according to neoclassical economists – who often cite Adam Smith, grandfather of modern economics. In his 1776 book, *An Inquiry into the Nature and Causes of the Wealth of Nations*, Smith observed:

> It is not from the benevolence of the butcher, the brewer, or the baker that we expect our dinner, but from their regard to their own self-interest. We address ourselves not to their humanity but to their self-love, and never talk to them of our own necessities, but of their advantages . . . Nor is it always the worse for the society that it was not part of it. By pursuing his own interest [a person] frequently promotes that of the society more effectually than when he really intends to promote it. I have never known much good done by those who affected to trade for the public good. It is an affectation, indeed, not very common amongst merchants, and very few words need be employed in dissuading them from it . . .[5]

Adam Smith's insight about how we help others by helping ourselves is backed up in modern economics using a collection of assumptions and relatively simple mathematical proofs. How? Smith uses his famous metaphor of the Invisible Hand to capture how, across a marketplace, everyone's selfish choices are coordinated, via shifting prices, to achieve what is best for everyone in the market as a whole. The price mechanism is neither tangible nor concrete. We cannot see lots of other people wanting to buy and sell stuff, but prices rise and fall to reflect the shifting balance of those who want to buy versus those who want to sell. In these anonymous marketplaces, we will gain nothing from attempting to second-guess others' choices. Our best strategy is selfishly to focus on ourselves and let the Invisible Hand of the price mechanism coordinate all our choices so that the prices paid reflect each person's willingness to buy or sell.

Of course, there are all sorts of problems with this account of price movements. Economists are often accused of promulgating a perspective on human behaviour that is excessively stark and unrealistic. And Adam Smith's views on our social lives were much more complex and nuanced than some might imagine just from reading selective quotations. More generally, economists make unrealistic assumptions to abstract from the complexity of the real world. Some economists argue that such assumptions help us to simplify and so capture the essence of human behaviour. The complexity is particularly significant when people are interacting by copying and herding. So, economists bring *Homo economicus* into their models, not because they believe real people operate in such a logical and mathematical way, but because it simplifies the analysis, especially when economists are investigating numerous, complex interactions between large numbers of people.

We can see this most clearly in a macroeconomy – essentially a crowd of crowds. Capturing group and herding behaviour within a small group is hard enough, but macroeconomists face an even greater challenge. To capture myriad interactions between people across an economy, macroeconomists have conventionally categorised different breeds of *Homo economicus* using assumptions about *representative agents*. These representative agents capture the stereotypical behaviour of key decision-makers in the economy, and they include representative worker-consumers and representative producer-employers. In a conventional macroeconomist's account, the representative worker-consumer makes a decision about how much they want to work, balancing the wages they can spend on consuming the things they enjoy against the discomfort and inconvenience of working. Workers have a symbiotic relationship with the representative employer-producers, who maximise their profits by employing workers at the lowest feasible cost so as to produce all the things that the worker-consumers want to consume. If these different groups of representative agents are identical and behaving in the same way, then economists can more easily analyse macroeconomic phenomena. They can add together the representative agents' choices via relatively easy arithmetic calculations.

What has this got to do with herding? Economic models of herding bring the same types of representative agents into their technical, mathematical analysis of how and why people copy others around them. In the case of self-interested herding, each member of the herd is rationally and individualistically pursuing their own self-interest – asking themselves 'What do I gain if I join?' Benefits may be immediate if we are able to make better choices for ourselves by following other people's good ideas and choices. Other benefits may be indirect and delayed. Sometimes we join a group because we believe that cooperating with others will deliver us long-term rewards. Many long-term collaborations and relationships involve patiently incurring costs in the beginning to ensure larger rewards in the end. Whether leading to short- or long-term gains, these choices are conscious and cognitively driven, inspired by a spirit of cooperation and collaboration but in ways that are consistent with self-interest and rational choice.

Social learning

Another feature of economists' representative agents is that they are super-rational and clever with information, using complex mathematical rules to process information efficiently. Herding is one manifestation of this clever information-gathering strategy. Rational herders identify strategies to minimise the costs to them of searching for information to guide their choices. They do this by balancing their *private information* against their *social information*. Our private information includes all the things we know that others cannot know we know. It is the information we have that other people cannot see because it is inherently unobservable and we cannot read each other's minds. Social information is the information we gather from observing other people's actions, and we use it to *infer* what caused others to act as they did. Just as other people cannot know what we know just by looking at what we do, so we cannot know for sure what they know just by watching them. But, by observing the choices they make, we can infer something about their incentives, motivations and intent. In the context of herding, we may conclude – though not always correctly – that the choices of others reflect underlying knowledge or expertise that we don't have. Often, we will not know, and may never find out, whether their knowledge is truly superior to ours. Consider the example of the vote at the meeting discussed at the start of this chapter. Voting in favour of a motion because others are doing so is consistent with rational choice if our vote is based on a rational calculation that the colleagues we are copying are better informed than us, and so we would do well to emulate them. Social information enables social learning: by observing other people's choices, and the rewards or costs those choices confer, we can learn about what is best for ourselves.[6] It is particularly important in situations where information is scarce and uncertainty is endemic. Why? Because, when we know very little, what we observe in others' behaviour and choices might be the best information we have.

Information cascades

In the early 1990s, economists started to develop a keen interest in the phenomenon of herding. They developed a range of theories and experiments to explore models of rational herding based around principles of social learning. They focused on explaining how we rationally balance social and private information, and how self-interested herding unfolds as a consequence. Pioneering studies of herding were developed by a team of economists including Sushil Bikhchandani, David Hirshleifer and Ivo Welch, based at the University of California.[7] They described self-interested herding as a sequential social learning process, with each person balancing what they already know against what they see others doing. The herd grows when each individual discounts what they privately know themselves and instead decides to follow the person in front of them. Bikhchandani, Hirshleifer and Welch use the powerful metaphor of what they call an *information cascade* to describe this herding process. One person makes a choice, the next person observes them and decides to do the same. Then the next person observes them and does the same too, with more conviction because they have had a chance to watch two people decide, not just one. As more and more people copy more and more people ahead of them, the power of the herd's signal increases. Social information about other people's actions flows through a group, building momentum as the herd grows. In other words, social information cascades through the herd. Information cascades help us understand a wide range of fragile and unstable phenomena in our economy and society, including booms, crashes, fads and fashions.[8]

Independently, the MIT economist Abhijit Banerjee developed a similar model of herding, illustrated with the everyday example of choosing between two restaurants.[9] Imagine that Restaurant A is crowded while next door Restaurant B is empty. Why don't the customers move from one to the other? Banerjee explains this apparent anomaly as evidence of rational herding.[10] People have a private signal favouring Restaurant A – say, a restaurant review they have read, or a recommendation from a friend. They can also collect some social information – they can observe which restaurant the other people ahead of them have chosen. Sometimes this social information conflicts with the private signal: someone has a recommendation favouring Restaurant A but sees a long queue waiting for a table at Restaurant B. The queue may encourage them to disregard their private signal and choose the crowded restaurant instead.

Banerjee's restaurant problem can also illustrate how information cascades work in practice. Imagine you face a similar conundrum to that posed by Banerjee. You are choosing between two adjoining Mexican street-food restaurants, Amigo's and Benito's. Assume that you know that Amigo's has in the past been favoured by most people, and Benito's by not so many. So, at the start, the balance of probabilities is in favour of Amigo's. However, you have read a restaurant review praising Benito's for its delicious tacos, tostadas and enchiladas. The private information you have suggests that Benito's is better.

Let's imagine that you join a crowd of restaurant-goers outside the two restaurants, and each of you decides, one by one, which restaurant to eat in. Your choice is complicated by the fact that the people waiting alongside you also have valuable private information. They are strangers, and so you don't know what they know or what might motivate them to choose one restaurant over the other. They may have read the same review that you read, praising Benito's. They may have read other reviews also raving about it. They may have heard from friends and family that Benito's is a great restaurant. So, even though Amigo's has been preferred in the past by most people, the unobservable private information suggests that the past preferences of the majority are unreliable. But no-one knows this because each person is deciding as an individual without knowing what others know. Adding to this confusion, let's assume that there is one person in the crowd who has contradictory private information: perhaps they are the only one to have read a biased online review from one of Amigo's friends suggesting that Amigo's is better. To make the problem particularly tricky, let's assume that that person gets to make their choice of restaurant first. They duly choose Amigo's.

Now it is your turn to decide. You have three pieces of information. The first is the publicly known information that most people have preferred Amigo's in the past, presumably for good reasons. The second is the social information you observe from seeing the first person choose Amigo's. The third is your private information: the restaurant review you have read recommending Benito's. This private information is consistent with the information held by all the other restaurant-goers except one. There are lots of different pieces of private information, but they all confirm that Benito's is better. So, your private information is, in fact, very reliable – but you don't know this because you can't observe the private information of those behind you waiting to choose. Nor can you infer anything from others' choices because you are the second person to choose. What should you do?

Let's say you choose to disregard your private information from the restaurant review and follow the first person in choosing Amigo's. The person behind you also knows that most people have favoured Amigo's in the past. Along with almost everyone else (aside from the person who decided before you), they have some private information suggesting that Benito's is better. But the person behind you sees both you and the first person choosing Amigo's. So, even though they also have information suggesting Benito's is better, they go along with the balance of probabilities and choose Amigo's too. And they choose it with more conviction than you chose it because they see two people ahead of them making the choice, whereas you saw only one. This is the information cascade and, as it takes hold, it feeds on itself. As more and more people choose Amigo's, more and more people are likely to make the same choice. They are doing so simply because others have chosen it, and not because there is any rich store of information underlying the choices they are observing. The herd streams through the doors of Amigo's, and not through those of unlucky Benito's, even though Benito's is the better restaurant.

An important point about restaurant queues specifically and information cascades more generally is that it is not necessarily irrational to follow others. It was not stupid for you and your fellow restaurant-goers to choose Amigo's over Benito's, even though these choices were generated within a fragile information cascade. The choices were logical and rational given the limited information available. To capture this logical nature of herding, Sushil Bikhchandani and his colleagues analysed information cascades using methods based around a mathematical theorem known as Bayes' rule, named after the eighteenth-century non-conformist minister and prototypical maths-geek the Reverend Thomas Bayes.[11]

Bayes' rule captures how we use different types of information to infer something from what we observe. We update our estimate of the probability of an event as new information comes along. We start with a *prior probability*, founded on all the information we have at a given moment. Then we learn something new, and we use this new information, together with our prior probability, to form a final estimate of the chances of an event. This final estimate is our *posterior probability*. The mathematics of Bayes' proof are complicated, but Bayes' rule has been applied widely, not only theoretically, to everyday problems by economists, mathematicians and statisticians. Some social scientists, psychologists and economists have also explored some of the ways that we use Bayesian-style reasoning, including when we herd. Economists like Bikchandani and his colleagues use Bayes' rule to explain how people adjust their probabilities when new social information comes along. Which restaurant is better? Which broadband deal should I choose? Will house prices go up or down? In answering any of these questions, other people's choices provide useful information, and we will use that together with any private information we already have.

Now we know more about how Bayesian reasoning works, let us return to the problem of reconciling conflicting evidence about the relative merits of Benito's and Amigo's. We start with a prior probability: most people preferred Amigo's in the past. Private information from a restaurant review we have read contradicts this. Then new information comes along in the form of social information about other people's restaurant choices. Using Bayesian reasoning, we update our estimate of the chances that one restaurant is better than the other to form a posterior probability. We reassess our initial judgement, deciding that the balance of private information and social information indicates that Amigo's is the restaurant to choose. We might reach the opposite conclusion if we could see everyone's private information – but we can't.

Herding games

If herding and information cascades are driven by people cleverly using Bayes' rule, then herding is not necessarily an irrational phenomenon. Nevertheless, whilst we have learnt that rational herding is a theoretical possibility, we have not established empirically that herding is rational. What is the evidence, either way? Across the social sciences, the answers are mixed. Some economists have collected evidence to suggest that herding is rational. But many other social scientists have collected evidence to suggest that it is not – as we shall see in the next chapter. Here, let's focus on the economists' evidence, and in subsequent chapters we will attempt to reconcile this with the conflicting evidence from other social scientists.

One piece of evidence comes from studying real-world restaurant queues. Behavioural economists Arthur Fishman and Uri Gneezy had a clever idea for a natural experiment to test for social learning about restaurant choices.[12] They recruited some research assistants to watch people choosing between two very similar fast food restaurants in an outdoor food court next to Bar Ilan University in Tel Aviv. They incorporated two observation periods into their study in order to capture how the impact of social influences shifted as people had more opportunity to learn for themselves which restaurants were better. As these restaurants were next door to a university, Fishman and Gneezy assumed that, at the beginning of the academic year, a larger proportion of customers would be new students (and so far less well informed about the restaurants' quality). So they observed one group of 1,324 customers in October 2009 (the beginning of Bar Ilan's academic year) and a second group of 1,153 customers around mid-April 2010 (the end of Bar Ilan's academic year).

Fishman and Gneezy discovered that there were big differences in the length of the queues of customers waiting for a table in the two restaurants. In October, the queues outside the crowded restaurant were much longer than those outside the emptier restaurant. By April, however, the queues were much more equal in length: whether the restaurant was crowded or empty was not making much difference to the queues' length. Fishman and Gneezy explained that social learning could explain the disparity. If the student customers had no prior knowledge and were inferring nothing from the choices of other customers, then they should have chosen randomly in October. The fact that they distributed themselves *unevenly*, joining long queues for the restaurant that was already popular and crowded, suggested that something else was driving them. Given that the only information available was the social information implicit in the choices of other customers, Fishman and Gneezy concluded that the queue was the trigger. Perhaps new students were using the social information conveyed by long queues as a signal of quality: a real-world example of an information cascade. By April, however, perhaps the students had had a chance to learn more for themselves about the two restaurants and so were less reliant on learning by observing others' choices, so the lengths of the two restaurant queues became much more similar.[13]

The American economists Charles Holt and Lisa Anderson, from the University of Virginia and the College of William & Mary respectively, explored the social learning, information-cascade hypothesis using controlled laboratory experiments. Holt is an experimental economist well known amongst economics lecturers for developing a wide range of engaging experiments, many of which are suitable for students to use in a classroom setting.[14] His experiments with Anderson were designed as a rigorous test of whether or not information cascades are consistent with Bayes' rule. Anderson and Holt's basic design has been widely replicated and refined in subsequent experimental studies, making it a very influential study for economists interested in herding.[15]

Anderson and Holt brought together seventy-two students to play a guessing game, with cash rewards for correct guesses. The students were shown two urns, Urn A and Urn B. Urn A contained two red balls and one black ball. Urn B contained two black balls and one red ball. Without the students seeing, the experimenters poured the contents of one of the urns into an unmarked urn. The students were then challenged to guess if this unmarked urn contained the contents of Urn A or Urn B.

To simulate an information cascade, the students did not guess all at once. They were asked to form a queue and guess one by one. They were given some extra pieces of information – some private, some social – to help them decide. The students got their private information from being invited to go up to the unmarked urn individually, pick out a ball, check its colour and then put it back, without letting any of the other students know the colour of the ball they had chosen. Each student then announced their guess of Urn A or Urn B to the group. One by one, the students were inferring something from the social information they were accumulating as they learnt about the other students' guesses.

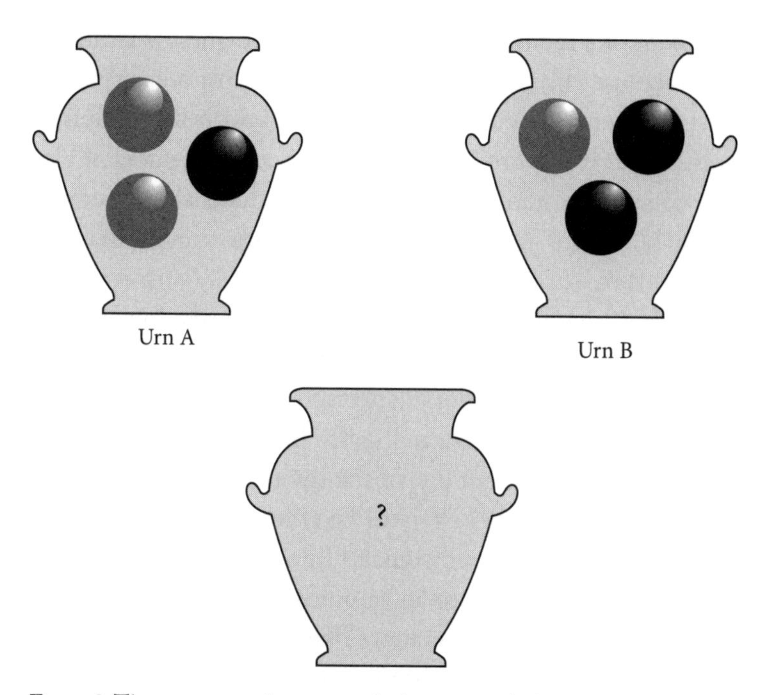

Figure 2. The urn game: players are asked to guess which urn's contents are in the unmarked urn: Urn A (2 red balls, 1 black ball) or Urn B (2 black balls, 1 red ball)?

Anderson and Holt postulated that the students were engaged in a process of Bayesian updating. Each student would form a prior probability based on what they knew at the outset. They updated this prior probability each time they heard another student's guess, and when they picked a ball themselves.

How does a Bayesian information cascade unfold in the urn experiment? Let us put ourselves in the shoes of the second student to decide, Bob. The first student, Alice, has already announced her guess – Urn A. Bob infers that this must be because she has selected a red ball, as there are more red balls than black balls in Urn A. Bob then draws a red ball from the unmarked urn. He now has two pieces of information: first, social information from Alice's guess of Urn A; second, private information from his own private selection of a red ball. Luckily for Bob, the guess is relatively easy because the social information and private information are consistent. He guesses Urn A. His guess is not definitely correct, but it is more likely to be correct than a guess of Urn B – which he would have no justification for making, because so far he has no *evidence* at all that the urn is more likely to be Urn B.

We can change the scenario to make it harder for Bob and to illustrate Bayesian principles. Let's assume it is Urn B, but that Alice's guess does not change: she guesses Urn A, so Bob infers that she did pick a red ball – by no means an impossible scenario given that one of the three balls in Urn B is red. But Bob picks a black ball. Now he is confused. What should he do, given these mixed signals? Should he go with Alice's guess of Urn A? Or should he guess Urn B, given that the black ball he has chosen is more likely to come from Urn B? If he guesses Urn A, he is discounting his private information – the evidence from his own eyes of a black ball. But if he guesses Urn B, then he is disregarding the information implicit in Alice's guess. For Bob, applying Bayes' rule could rationally justify either answer.

Let's assume that he decides to favour the social information from Alice and guesses Urn A. Then a Bayesian information cascade will start to build. The third student, Chris, draws his ball and perhaps again picks a black ball. Chris has three pieces of information. Alice has picked Urn A, and so has Bob: Chris assumes that this is because they have picked red balls. Chris, however, has picked a black ball – one piece of information that suggests Urn B, against the two inferences he makes from Alice and Bob's guesses of Urn A. The balance of evidence has shifted in favour of Urn A, even though that is not the right answer. If Chris is using Bayes' rule then the only conclusion he can reach is that he should guess Urn A. For Chris and all the students still waiting to guess, rationally that is the best guess they can make. The information cascade reinforces Alice's mistaken guess of Urn A. So, no student will win a cent if they are using Bayes' rule to decide. This information cascade has led the herd in completely the wrong direction and the pinch point was Bob's choice, when the guesses were on a knife-edge. If Bob had instead favoured his private information and correctly guessed Urn B, then all the students except Alice would have won money for correct guesses (and it would have turned into a very expensive experiment for the researchers).

Anderson and Holt analysed all the evidence from their experiment to assess whether the students were deciding in a way that was consistent with the Bayesian information cascade models described above. They found that information cascades unfolded in a way that was consistent with Bayes' rule in forty-one out of the fifty-six times when the private information and social information were inconsistent – that is, in the sort of situation Bob faced when he saw a black ball alongside inferring that Alice had seen a red ball.

What of the fifteen of the fifty-six times when the information cascades were *not* consistent with Bayes' rule? What explains those guesses? Were some students better at using Bayes' rule than others? Do the anomalous findings suggest that some people place different weights on private and social information? Could students have been using a simpler rule of thumb to decide, and this rule, just by coincidence, generated guesses that mimicked Bayesian guesses?[16] Anderson and Holt's experimental findings have been replicated across a wide range of other studies but not many have rigorously tested alternative hypotheses. Do most of us use Bayes' rule to process social information? Or do we use other tools to guide our choices? Economic theory does not answer these questions, and so we shall go beyond economics to explore some answers from other disciplines in the following chapters.

Is social learning good or bad?

From an economist's perspective, is following the herd rational or irrational?[17] If the herd goes in the wrong direction, then that is obviously bad: a large group of people have made the wrong choice.But even if the herd is on the right track, there will nonetheless be negative impacts because valuable private information is lost when people disregard it in the process of following a herd. We can use our restaurant example to illustrate the point. Once the information cascade favouring Amigo's takes hold it will continue until everyone has chosen Amigo's. At the end of this process, many pieces of useful, rich, privately held information will have been discarded by the herd. A negative, suboptimal outcome has emerged because individuals have favoured social information over important, useful but unobservable private information.[18] As individuals' private information is lost during herding, there are negative external consequences for the group – what economists call *negative externalities*. Restaurant-goers have forgone an opportunity to try Benito's and discover how good it is. If they had chosen it, Benito's would have justly benefited from increased takings and the buzz of popularity. Those enjoying Benito's might later have had a chance to share their good experiences with friends and family, and with others more widely via online reviews. There would have been many winners and only one loser (Amigo's) if the herd had headed in a different direction.

Perhaps counterintuitively, these negative consequences do not disappear just because the herd has identified the better path. A subtler point is that, even if the herd had headed in the right direction in choosing Benito's, private information would still have been lost and overwhelmed by social information. Imagine that one of the people who had some private information suggesting Benito's was better had been the first to choose which restaurant to eat in, thereby setting off the unfolding of an information cascade that ensured the herd made the right choice. The point is not so much about whether the herd does the right or wrong thing in the end, or whether each person has decided in a logically rational way. The problem is that rich stores of private information are lost via this mechanical Bayesian updating process.

We can illustrate the importance of private information if we change our restaurant scenario a little. Imagine that the first person to choose hasn't read a biased online review but instead has read a very recent review written just after Amigo's had sacked its cook and enticed Benito's brilliant chef away with a promise of better pay and working conditions. So, the good review for Benito's, read by us and most of the others waiting, was based on inaccurate, out-of-date information. The first person to choose had better private information, i.e. a bang-up-to-date and possibly more accurate review. Still, perhaps the brilliant chef will not do so well at Amigo's if Amigo's has other problems besides the cook they have just sacked – poor management practices, perhaps. Either way, a rich, diverse set of private information is helpful or, at the very least, might help each restaurant-goer to know that there isn't unanimous agreement about which restaurant is better. Any and all of this information is lost once the information cascade takes hold.[19]

So, self-interested herding driven by social learning can create distortions. Are other forms of self-interested herding less problematic? Some can be helpful for the group as well as the individual. To see how this works let us turn to some of the other economic incentives and motivations behind self-interested herding. There are strategic advantages when we copy others, linking to the benefits we gain by using herding as a form of signalling. Self-interested herding can be a means to build our reputations. Powerless individuals can gather together in a powerful herd. Herds are sometimes havens for safety.

Strategic advantages

The strategic advantages that we can accrue if we join a group or herd have been extensively explored by game theorists.[20] The basic idea is that a selfish individual can hook up with other selfish individuals and together, as a group, they can do much more than each person could do alone – for example, when hunting. In his 1755 masterpiece, *A Discourse on Inequality*, philosopher Jean-Jacques Rousseau used a 'stag hunt game' to illustrate how coalitions form for the benefit of each member.[21] Four hunters are deciding whether to hunt as individuals or to collaborate and hunt as a team. No one hunter can catch the stag alone because it is so big and fast. If they hunt as individuals the best they can hope for is to catch a hare. One hare is not even enough to feed a single family. A much better outcome would be for all four hunters to join forces and catch a stag together. A stag would be more than enough to feed four families, whereas a single hare would leave each family hungry. So the hunters form a coalition. Assuming the four hunters can negotiate an equitable division of their hunting spoils, then their coalition will prosper. The benefits of working together for the individual members of the coalition are greater than if each of them had hunted alone. It is in the individual's self-interest to join the hunt: everyone's a winner (except the stag).

Groups of self-interested individuals do not always deliver a good collaborative outcome, however. When people work together they interact, and so selfish individuals can affect the actions and performance of the group as a whole. When outputs and rewards are shared in a team, the individual team member may have incentives to shirk and free-ride on the efforts of others. Self-interested individuals will subvert the efforts of the team, unless everyone's incentives are somehow aligned. This insight about strategic advantage parallels economists' models of rational herding as a response to the extra benefits that can come from copying other people's choices. The most common example is the additional payoffs that accrue in financial markets when a series of financial traders are buying into a rising market, each helping an asset's price to rise and thus benefiting the whole herd of traders. We shall explore these related financial herding phenomena in chapter 6.

Signalling

Another manifestation of self-interested herding is the copying behaviours we use as signals to others around us.[22]

For example, unconventional behaviour can be used as a signal of authenticity and commitment to groups defined by their rebellion against society's norms. Twentieth-century youth subcultures – from mods and rockers to punks and goths – show how signalling reinforces our sense of identity. In a world of imperfect information and limited trust, we are vulnerable to exploitation by those who can pretend to be what they are not. Behaviours that might seem contrarian to the world at large are crucial signals we send to important subgroups with which we identify; those groups are more likely to trust us if we resemble them, and we are more likely to trust them – to our mutual benefit.

We will explore the perspective of the group in more detail in the next chapter, but some economists have explained how and why we form an identity using the standard economics focus on balancing benefits against costs. In this way, economists George Akerlof and Rachel Kranton explain how we use signals to build identity. Actions that might seem anomalous to outsiders have payoffs for members of a group because they help a person to build their sense of identity with the groups they join. Identity and belonging increase people's satisfaction and so they will be prepared to incur physical and economic costs in acquiring physical markers that accentuate their sense of belonging to a particular group.[23] When and how is it economically rational to signal our identification with others through ostensibly costly and painful actions, such as tattoos and piercings? These seem like maverick actions to outsiders, but make much more sense to others with whom we identify. And the costlier the actions, the better, because more costly signals are more credible. We would not incur such large costs – whether physical, psychological or monetary – if we were not sincere.

The political scientist Henry Farrell has explored unconventional behaviour in the seemingly unlikely context of the personal grooming of hipsters – analysing a debate between economist Paul Krugman and journalist Ezra Klein about the purpose of tattoos versus topknots.[24] A hipster's topknot is not a costly action – it is easy to do and to remove – so, in strategic terms, members of a group of top-knotted hipsters will not interpret your top-knot as a credible signal of a strong affinity. If you want to send a costly – and therefore more credible – signal to other rebels and minority groups that you are sincere about joining, then a tattoo is more convincing because it is not 'cheap talk'. You show others that you are serious by going through painful actions at significant personal cost to yourself. Farrell links this to sociologist Diego Gambetta's insights in *Codes of the Underworld*, his study of how criminals communicate with each other: 'Erefaan's face is covered in tattoos. "Spit on my grave" is tattooed across his forehead; "I hate you, Mum" etched on his left cheek.' Permanent facial tattoos are outwardly unconventional actions but they are costly and therefore a much more credible signal of commitment, essential to acceptance by specific rebel groups. 'The tattoos are an expression of loyalty . . . you are marked, indelibly, for life. Facial tattoos are the ultimate abandonment of all hope of a life outside.'[25]

Initiation rites and frat house 'hazing' serve similar purposes. On the surface, these behaviours seem perverse and ultimately contrarian, but if people are using unconventional behaviours as a way to build alliances with groups whose identity they would like to share, then this makes much more sense. Defying social norms is sometimes consistent with self-interested herding. If a self-interested rebel has much to gain personally from joining a group of like-minded rebels, then it pays for them to incur costs to imitate the other copycats within the rebel herd.

Signalling is not just directed at the groups we wish to join. We also signal our virtues as well as our status. Car choices are a classic example of the ways in which we use different signals. A person who buys a Maserati is signalling status, and it works because they are imitating others before them who have signalled status in the same way. An environmentalist who buys no car at all may be signalling to other environmentalists that they share with them a virtuous regard for the environment.[26] This social signalling operates at all levels of society. One research study explored the behaviour of poor families lacking the money to pay for basic foodstuffs. When they were given additional income, they spent it on consumer goods such as TVs even though their families were malnourished.[27] This is not necessarily irrational. We live and work in social groups, and if we are to survive and prosper in these groups we need to attract the respect of the rest of the herd. If others are impressed by our standard of living, then our lives might be easier.[28]

Conformity also has a value that connects with our social rankings. The economist B. Douglas Bernheim has explored the ways in which status encourages conformity with the group from the perspective of a selfish individual maximising their own utility. At a social level, status is important and improves people's satisfaction. Being ostracised for departing from conventions and social norms will threaten our status, and so fads and customs will persist for longer than they are useful. Self-interested copycats recognise the negative consequences of deviations from social norms and, conscious of what they will suffer from rebellion, they conform and follow a herd.[29]

Reputation

Signalling connects closely with reputation, although signalling is a more ephemeral phenomenon and reputation is something we are all keen to build over time. A good reputation has value, both tangibly and intangibly. Reputations are more vulnerable in today's digital age. We might hesitate to reveal our Saturday-night excesses on Facebook and other social media sites were we to consider the potential impacts on our future reputations, for example when looking for a job. We take fewer risks with our reputation when we are following others around us. The economist John Maynard Keynes is famous for observing this: 'Worldly wisdom teaches that it is better for reputation to fail conventionally than to succeed unconventionally.'[30] In the modern world, rogue traders are an example of the vulnerability of a reputation built on contrarian choices. Spectacular gains can be made when a trader bids against financial market conventions. But when the crowd is right and the contrarian is wrong, reputation cannot so easily be saved. Contrarian traders cannot simply defend themselves by arguing that their mistake was a common one.

In the business world, firms that value their good reputation can be steered towards better behaviours overall, and firms will follow other firms in adopting best practices. Business corporations' preoccupation with fairness and legal and ethical requirements is not driven by altruism, however, but rather reflects an enlightened self-interest. Corporate management teams realise that their business is more likely to survive if they have a good reputation. An example of how these influences might gain traction is in firms' approaches to environmental policy. Corporations can build market share by signalling to the world that they are 'good' and that consumers should therefore support their products. When the US Chamber of Commerce opposed climate-change mitigation policies in 2009, a series of resignations by executives from Apple, Nike, Pacific Gas and Electric, Exelon and PNM Resources followed.[31] Conversely, companies have been vilified for not paying enough taxes. If a herd of consumers is large enough, it can effectively pressure companies into wide-scale changes in commercial practices and partnerships. In the aftermath of numerous school shootings, finally catalysed by those in Florida in February 2018, a variegated herd of businesses – from car hire businesses Hertz and Avis through to key-maker Chubb and the First National Bank of Omaha – acted in defiance of the politically powerful National Rifle Association. They removed various deals and privileges for NRA customers, a response to widespread pressure from anti-gun protesters.[32]

In the context of environmental strategies, corporations build their reputations through their corporate social responsibility programmes. These often include commitments around environmental responsibility, partly as a response to consumer pressure and corporate concerns about keeping their customers happy by behaving in ways that consumers think is fair and principled. The corporation's wider reputation, including with investors and competitors, will also play a role.[33] So, reputational concerns can encourage corporations into more environmentally sustainable and innovative methods of production. Corporations may compete for reputation – especially if information about firms' environmental records is made more easily available to the public, as with environmental blacklists. One example is the Toxics Release Inventory in the US, which acted as a form of social signal. It helped consumers learn about different firms' environmental records so that they could discriminate in favour of those with good environmental practices. This links with the push for a 'Greenhouse Gas Inventory', as advocated by Richard Thaler and Cass Sunstein in their bestselling book *Nudge*. Thaler and Sunstein explain that policymakers can use social influences to 'nudge' consumers and firms in a better direction. The essence of nudges is that they are little pushes in the right direction. They are a form of what Thaler and Sunstein call libertarian paternalism. They are libertarian in that individuals are still able to choose for themselves.

Nudges are not sanctions and they do not impose direct costs on the individual, as taxes would. People can ignore the nudge if they want to. But nudges are also paternalistic because they are designed and implemented by policymakers to achieve publicly desirable outcomes. If well designed, people will use nudges as a signal helping them to decide what is the best strategy for them and others around them. Social nudges are a common form of policy used in the energy and environmental sector – and we shall see a few examples throughout this book. If significant emitters were obliged to disclose emissions levels via a Greenhouse Gas Inventory, then they would be revealing information to their customers. Benefits ensue, not only in terms of making relevant information more transparent for environmental regulators, but also via consumer pressure. Consumers concerned about climate change will have information about the worst emitters and will pressure those firms into reducing emissions. In an age when online social media are ubiquitous and powerful, bad publicity spreads quickly. It damages relationships with competitors and investors, as well as with customers.[34] So the self-interested directors and managers of commercial firms have reasons to imitate other firms' best practices if this helps them to build their corporate reputations.

Power and safety

Another motivation for self-interested herding is the power that the individual can gain from joining a group. Collective action is, in many important contexts, more powerful than individual action.[35] Groups can give individuals security, especially when they provide safety in numbers. For example, the herd protects pedestrians when they are crossing busy roads. If you have ever been in an overcrowded city such as

Jakarta, especially as a stranger, you may have been disconcerted at the thought of crossing congested main roads jammed with cars and motorbikes. The less you know about a city the harder it is to resolve your problem because your trust in local drivers may be limited, or you may know less about the city's traffic rules and driving conventions. What is the best strategy for getting where you want to go? The quickest way might be to move with a group of locals because you are learning by observing the local pedestrians' habits. You will also enjoy safety and shelter from harm by belonging to a larger group. A car is far more likely to run over a lone pedestrian than a crowd. A negative consequence of this grouping behaviour is that an extremist contrarian wanting to attack a crowd violently can succeed more easily when we herd together, with severe consequences. The truck and van attacks perpetrated by terrorists across Europe and in New York in 2017 depressingly illustrate that the crowd is not always a safe place to be.

Beyond physical safety, in our civil lives there are corollaries of the advantages we gain from joining groups and herds. Groups have much more political clout and influence than individuals. With class action suits, for example, otherwise powerless individuals can leverage group power to get justice for themselves. Many class action suits relate to illnesses and deaths caused by harmful chemicals. One example is the case of the 'fen-phen' drug (a diet pill made by mixing the appetite suppressant fenfluramine and the stimulant amphetamine phentermine). These were marketed by the American Home Products Corporation (now Wyeth) and had been prescribed by a range of medical practitioners before being withdrawn by the US Food and Drug Administration in 1997, when scientists found that the use of fen-phen was associated with side-effects including hypertension and heart valve damage. The thousands of users of the diet pills who had suffered the side-effects came together and, in 1999, the American Home Products Corporation agreed to pay the plaintiffs a total of $3.75 billion. This was not the only legal action, and Wyeth was later forced to set aside $16.6 billion to cover its fen-phen liabilities.[36] The plaintiffs' choice to join with others in the legal action was a rational, self-interested choice from each individual plaintiff's perspective. Self-interested herding alongside others suffering similar disadvantages gave power to the individual plaintiffs. They would have had no power at all if they had acted alone.

A key limitation of the economic approaches to copycats and contrarians highlighted in this chapter is that they are founded on a fundamental belief in individuals' capacity for logical, rational decision-making. The Bayesian calculations forming the foundation of information cascade models are simpler than the complex mathematical calculations embedded in many economic models. Even so, Bayesian models cannot capture complex sociopsychological influences. Bikhchandani and his colleagues have acknowledged that, although their economic model of herding as an information cascade captures the fragility of herding in a simple setting, it cannot explain why mass behaviour in the real world is fragile. Their models cannot explain why changes in social and political attitudes are sometimes so unstable, for example in the context of changing attitudes towards lifestyle choices such as cohabitation, sexuality, communism and addiction.[37] To be capable of Bayesian reasoning humans would have to have relatively high levels of numeracy and logical capacity, when in reality most of us do not think in such sophisticated ways.[38]

Reflecting on all these different explanations for self-interested herding, economists tend to rely on the idea that humans are good at mathematical reasoning. But have humans really evolved the ability to effectively apply high levels of numeracy and sophisticated probabilistic reasoning? The capacity for complex and abstract mathematical calculation would not obviously have bestowed evolutionary advantage in hunter-gatherer settings, and so it is hard to imagine where such high levels of numeracy might come from. Another problem with economic models of self-interested herding is that they tend to start from the perspective of the individual decision-makers and the incentives and motivations driving them. Yet herding may be a product of forces not easy to understand from an individual's perspective. What is rational for the group is not necessarily rational for the individual, and vice versa.

Moving beyond economics, other social sciences have developed a wider understanding of the social influences driving our behaviour. We are susceptible not only to the less obviously rational elements associated with emotions and personality traits, but also to losing our individual identities when forming part of a group or herd with a powerful identity of its own. All these insights can help us to understand herding as a collective phenomenon, explicable in terms of sociological and psychological forces, as we shall see in the next chapter.

2

Mob psychology

How and why would a group of close on 900 people collectively decide to collaborate in a mass murder-suicide pact? These were compelling questions in the aftermath of a terrifying massacre which took place in Guyana in 1978. Jim Jones, the founder and self-proclaimed 'Father' of the Peoples [*sic*] Temple of the Disciples of Christ, persuaded the members of his cult first to assassinate an American congressman, some journalists and a cult defector, and then to turn the metaphorical gun on themselves. Parents poisoned their children with cyanide-laced fruit drinks, and then killed themselves with a communally produced cocktail of cyanide and sedatives. Jim Jones shot himself on the same day.[1]

Most of us would find it hard to imagine how individuals could be manipulated into perpetrating such extreme and violent acts en masse. Jones had founded the Peoples Temple in 1955 in Indiana, blending Christian and socialist principles to further the cause of communism. The cult grew and moved to California, but in the early 1970s became the target of a series of exposés documenting abuse and exploitation within the cult. In 1974, Jones left to found 'Jonestown', seemingly as a socialist community agricultural project. He was joined by many members of the cult, yet, just four years later, they cut short their new lives in the 'revolutionary' self-slaughter – ostensibly voluntarily.

Why did so many otherwise conventional and law-abiding individuals allow themselves to be manipulated by one man? That is a question asked in relation not only to horrific isolated instances of violence such as the Jonestown massacre, but also more widely, across history. In the context of atrocities committed before and during the Second World War, many social scientists have hypothesised how and why large numbers of ordinary people not only stood by as passive observers, but also actively participated in the atrocities perpetrated during the Holocaust. Nor are such extreme levels of prejudice and violence a historical anomaly. As explored in later chapters, other studies in social psychology look at destructive, violent behaviours driven by social influences – specifically many people's unhesitating tendency to obey authority figures. Otherwise ordinary people can be encouraged by their leaders to commit cruel acts including administering extreme electric shocks and other forms of inhumane treatment.[2] These behaviours are all but impossible to explain using standard economic models in which people sensibly herd together as rational, self-contained and selfish individuals. Actual human experience is much messier, and abstract economic models are not well designed to describe all the real world's social and psychological complexities. In this chapter, we shall go beyond the ordered world of economics to explore insights about copycats and contrarians from the other social sciences, focusing on social psychology and sociology.

Collective herding and the wisdom of crowds

In the previous chapter, we saw how economists analyse herding as a clever strategy. Self-interested herding may create problems for groups, economies and societies at large; but from each individual's economic perspective, following others is often a sensible strategy. A quite different type of herding is *collective herding*. Collective herding is not about the wants and needs of self-interested individuals. It is about the motivations and goals driving the group as a whole. Groups often form their own independent entity in a way which is impossible to explain from the perspective of a single individual.

Although the foundations of self-interested herding and collective herding are very different, there are some resonances between them. Some perspectives on collective herding explain how the whole group functions as if it were a rational individual, and this is captured in the literature on the wisdom of crowds.[3] Individuals grouping together can sometimes come up with better answers than if they are all deciding separately and independently. The inspiration for the wisdom of crowds concept comes from the eighteenth-century French mathematician and philosopher Nicolas de Condorcet, and his analysis forms the basis for what is now known as Condorcet's jury principle. It is often applied to juries, a real-world example of how we place our hopes in the wisdom of a collective judgement.[4] But Condorcet's original analysis was not about juries at all. It was a highly abstract mathematical proof. Condorcet started his theory with a pair of decision-makers, each of whom is slightly *more* likely than not to know the right answer: Condorcet assumed that the probability that each decision-maker is correct is a little greater than ½. He then analysed what happened as other decision-makers were included in the decision-making.

Condorcet's mathematics showed that the chances of the group being correct increases and increases as more and more decision-makers join the initial pair. Eventually, as the pair grows from a group into a crowd, then the probability that they will, collectively, identify the right answer approaches 1. If the crowd is infinitely large, then it will almost certainly be correct. This seems like a great result – until we consider the opposite. Condorcet's mathematics also showed that if each individual decision-maker is slightly *less* likely than not to know the right answer – if their probability of being correct is just a little less than ½ – then the collective answer does not look so smart. As the pair of wrong-headed decision-makers grows into a crowd, then the probability that the crowd will, collectively, identify the right answer approaches 0. Under this second scenario, an infinitely large crowd will almost certainly be wrong.

The real-world question is: how can we ensure that our crowd includes people who, as individuals, are more likely to be correct than not? The American psychologists David Budescu and Eva Chen outline some strategies for leveraging wise crowds to improve collective decision-making. How can a crowd be designed to ensure that Condorcet's conditions for the wisdom of crowds are met? A simple solution is to exclude all poor performers: just omit the judgements of those who have a record of being wrong more often than they are right. Budescu and Chen support their hypothesis by analysing data from the Forecasting ACE (Aggregate Contingent Estimation) project.[5] This website collects together judgements from volunteer forecasters known as 'judges'. The judges do not have to be experts in any conventional sense. They are asked to forecast a range of events from economics through to politics, health and technology. Budescu and Chen collected and analysed data from the ACE website to assess the performance of 1,233 judges forecasting 104 events between July 2010 and January 2012. They scored the judges according to the accuracy of their predictions. By identifying the best contributors within the crowd and eliminating those whose forecasts were wrong more often than the average forecaster, Budescu and Chen showed that their selection method increased the accuracy of predictions by approximately 28 per cent.[6]

Embedding the wisdom of crowds idea more generally into real-world decision-making is problematic, however. Budescu and Chen selected the better judges on the basis of the accuracy of past forecasts. But in a world that is profoundly uncertain we cannot easily devise objective benchmarks against which we can judge who is getting it right and who is getting it wrong. A further theoretical problem with Condorcet's jury principle is its starting assumption that all the individuals' initial judgements are completely independent and uncorrelated – a testing assumption, especially given the human tendency to follow others. Opinions are more often correlated than independent. There may be a number of reasons for correlated opinions. Experts and others may share a belief in an established paradigm (e.g. 'the world is flat'). People who share identities may agree with each other even when there is little objective reason to do so. Biases in our thinking may lead us to agree with others when objective evidence suggests we should not.

Le Bon's psychological crowds

Condorcet's wisdom of crowds is a simple mathematical analysis that abstracts from the complexities of human psychology. It bypasses the important psychological drivers in our social lives: personality and emotions. If we have a conformist personality, then we will move with groups, herds and crowds more often. If we are curmudgeons, we will feel more inclined to rebel. Emotions are important too.[7] We join herds when we fear for our safety, or when we are anxious we might make the wrong choice. We join crowds to feel happy – at concerts, parties and parades. And emotions and personality will come together in driving our choices. Personality traits predispose us towards specific emotions. In turn, these emotions will determine whether we are inclined to join in or to go it alone. An introverted, anxious person might join a crowd if they feel threatened, but they will be less inclined to attend a big, loud party.

Gustave Le Bon was one of the early pioneers in the study of how our copycat psychology unfolds in herds and crowds, and his 1895 work *The Crowd: A Study of the Popular Mind* endures as a seminal analysis of crowd psychology. Le Bon was a French medical doctor who developed wide-ranging interests across the social sciences, particularly sociology and psychology. His fascination with mob psychology was driven by his curiosity about how crowds form around specific causes, and *The Crowd* draws strong parallels between the psychology of the crowd and political movements. This was a reflection of the instability of his times. Le Bon was born in 1841 and was a child during the 1848 'Year of Revolution', a year of great political significance marking a turning point in Western democracy. With increasing demands for new democratic institutions to replace old feudal structures, uprisings started in France and soon spread to many other European countries and beyond. As an adult, Le Bon was living in Paris during the brief, revolutionary government of the Paris Commune in 1871. In response to the violence he observed he developed a conservative attitude towards political uprisings. In his accounts of mob psychology he presents a dystopian view of the impacts of collective political action. Even so, his psychological work was politically influential. Jaap van Ginneken, a Dutch psychologist and former activist and journalist, observes that even though Le Bon's ideas were largely derivative they remained influential with a wide range of the twentieth century's political leaders (good and bad), from Theodore Roosevelt to Adolf Hitler.[8]

Le Bon was inspired by the ideas of French sociologist Jean-Gabriel De Tarde, who had argued that we are driven by conscious and unconscious motivations to imitate each other.[9] Building on Tarde's insight that imitation is one thing that is fundamental to our social interactions, Le Bon describes two very different sorts of crowds – what he called *organised crowds* and *psychological crowds*.[10] An organised crowd is a collection of individuals coincidentally gathered in one place – just a group of ordinary people going about their business in an ordinary way, with no obvious common purpose. Organised crowds may be large, but they are benign. Sometimes, however, organised crowds are transformed into Le Bon's psychological crowds, or what we might call a mob. Mobs are fundamentally different from organised crowds because they form a sinister identity of their own that cannot be explained from the perspective of any individual mob-member. Each individual loses their personality and sense of personal identity.[11] Each individual's intelligence is swamped, and so the mob is characterised by a lower degree of intelligence than the individuals within it:

> however like or unlike [are the individuals'] mode of life, their occupations, their character, or their intelligence, the fact they have been transformed into a crowd puts them in possession of a sort of collective mind which makes them feel, think, and act in a manner quite different from [what each] individual . . . would feel, think and act were he in a state of isolation . . . the intellectual aptitudes of the individuals, and in consequence their individuality, are weakened . . .[12]

Mob behaviour is impetuous. Instincts are unrestrained. For each person, their 'conscious personality vanishes'. Someone who might usually be sensible, logical and calm becomes wild and unruly. They become much more suggestible. It is as if the mob is exerting a hypnotic influence on its constituent members. Another famous Victorian writer on crowds and mobs, Charles Mackay, mirrored Le Bon's insights about how we lose reason in herds, observing: 'Men, it has been well said, think in herds; it will be seen that they go mad in herds, while they only recover their senses slowly, and one by one.'[13]

Le Bon's description of the mob is colourful and engaging, but what can he tell us about how to understand and analyse collective herding? One of his key lessons is that we cannot start by assuming that mobs are a simple aggregation of individuals. Mob members are not engaging in self-interested herding. The mob is driven by forces that are hard, if not impossible, to explain as the product of individual motivations and incentives.

Freud on belonging

If not self-interest, what does encourage us to join a group or mob? We cannot easily explain this solely in terms of the logical and tangible incentives and motivations that are the focus of economic analysis. Joining the herd gives us an ineffable sense of psychological satisfaction. Sigmund Freud, the father of psychoanalysis, developed some early insights about how our relationships with others affect our psychological lives, including the urges and instincts that propel us to join groups and herds.[14] Freud's analysis focuses on the roles played by our unconscious in shaping our feelings and choices. In *The Pleasure Principle* (1920), he argued that our personalities are prone to conflicts between our life instinct (Eros) and our death instinct (Thanatos). These connect with unconscious facets of our personality – the id, ego and superego, as Freud sets out in his 1923 masterpiece *The Ego and the Id*. Freud's analysis of the conscious and unconscious forces driving people's actions suggests that personalities are not formed as one homogenous whole. Psychological forces operating below the level of our consciousness are driving all our decisions, including our copycat choices. With self-interested herding, perhaps the more rational and deliberative ego is in control. With collective herding, perhaps the more instinctive and less rational id takes over.

Freud directly applied some of his insights to the analysis of mobs and crowds, reflecting his interest in the political psychology of mass movements. In *Group Psychology and the Analysis of the Ego* (1921) and *Civilization and Its Discontents* (1929), he developed Le Bon's idea that individual personalities are lost when we seek security within groups. Paralleling Le Bon's distinction between organised crowds and psychological crowds, Freud distinguished 'organised' and 'artificial' groups from 'common groups' – a corollary of Le Bon's psychological crowds. People become more susceptible to communal emotions and instincts when they join a common group. They lose independence, initiative and their sense of individuality. Their identification with the group overwhelms their own selves. Freud took on the idea that herd instincts are innate – a notion that British neurosurgeon Wilfred Trotter had developed in his popular book about herd instincts.[15] Trotter argued that herd instinct is a primary instinct, to be grouped with fundamental urges associated with self-preservation, nutrition and sex. Freud countered, arguing that our need to belong to a group has its origins in family relationships. All our drives to join groups and herds reflect our unconscious need to belong. In our unconscious minds, opposing the herd is as bad as separating from it, and separation generates extreme anxiety.

Drawing on Trotter's observation that people feel incomplete when they are alone, Freud argued that this anxiety parallels a similar fear in small children. According to Freud, the roots of this separation anxiety, and of social instincts more generally, lie in children's attachments to their parents. A child with a new sibling feels jealous but realises that their jealousy will poison their relationship with their parents. They sublimate their jealous feelings and replace them with familial feelings for their siblings. The child forms an affinity with their sibling to reconcile their conflict between jealousy and attachment to their parents. Freud argues that, as adults, this childhood conflict is generalised in our social feelings towards other adults around us. We reverse our hostility towards others and replace it with a more positive sense of a tie with others. So, perhaps ironically, envy leads us to identify with our rivals. This forms the basis for *Gemeingeist*, or 'group spirit'. Freud illustrates this with an example of fan behaviour:

> We have only to think of the troop . . . in love in an enthusiastically sentimental way, who crowd round a singer or pianist after his performance. It would certainly be easy for each of them to be jealous of the rest; but, in the face of their numbers and the consequent impossibility of their reaching the aim of their love, they renounce it, and, instead of pulling out one another's hair, they act as a united group, do homage to the hero of the occasion with their common actions, and would probably be glad to have a share of *his* flowing locks. Originally rivals, they have succeeded in identifying themselves with one another by means of a similar love for the same object . . .[16]

For these unconscious conflicts to work, all the followers in a herd must be equals.[17] Again, for Freud, this parallels childhood experience. For the children's jealousy to be held in check no one child within the family can be favoured, and Freud argues that this forms the roots of our preoccupation with equality within the herd.

Gestalt psychology and psychosociology

Freud's insights inspired other psychoanalysts and psychologists to explore the nature of groups and herds. We have emphasised already that we can only understand collective herding if we understand the mob as an entity with its own identity, an identity that is substantively different from the separate identities of the individuals in the mob. Groups, crowds and mobs cannot be understood by simply adding together the self-interested choices of the individuals within them, as economists tend to do. In his treatise *Metaphysics*, the Greek philosopher Aristotle observed, 'the totality is not, as it were, a mere heap, but the whole is something besides the parts'.[18] This idea is exemplified in the assertion from Gestaltian Kurt Koffka that 'the whole is something else than the sum of the parts'.[19] This essential principle of Gestalt psychology was originally applied in the context of visual perception. When we look at a photo we do not see a mass of pixels and dots. We see an image of something that is quite different in nature than the physical object. Optical illusions work on the idea that our perception changes as we shift our perspective. This idea of the whole being something other than the sum of its parts also links to group phenomena: like the photo, the group has a nature and identity of its own which we cannot understand just by looking at the individual group members as if they are separate pixels.

Wilhelm Reich, a psychoanalyst and pupil of Sigmund Freud, developed Gestalt principles in the context of the lesser-known social science of psychosociology. Reich was born in 1897 and, like Gustave Le Bon, was interested in the mass psychology of political movements, including the rise of fascism in the early twentieth century.[20] Reich aimed to bring together insights from political science and psychoanalysis. He argued that the structure of our characters develops as a product of social institutions and processes.[21] He believed that mental illness is not just about a person's character, as Freud would argue. Rather, he asserted that it also reflects the domestic and socioeconomic conditions in which people lived, and he drew on Marx's insights in developing this idea.

Echoing Le Bon, Reich argued that social groups influence us as individuals. Groups make us into something more than our independent selves. In groups, we are driven by the goals and desires of the whole group, and not by the interests of the individuals within it. Groups and individuals evolve from the influence of the other, reflecting tensions and conflicts between the two. The group changes the individual and the individual changes the group. The Jonestown massacre illustrates these interplays and feedbacks. The Peoples Temple changed its individual members: each joined the cult as ordinary Christians, a decision which transformed their lives, identities and destinies. Perhaps less obviously, the members also changed the cult. If Jim Jones had not been able to persuade anyone other than his close confidantes to join, then the Peoples Temple would probably have been forgotten. With so many individuals prepared to join it and to sacrifice so much to defend it, the cult's nature and identity changed. The Peoples Temple would have had neither power nor influence without its people. As the cult transformed the cultists so the cultists transformed the cult.

Some of Reich's ideas parallel similar analyses in economic psychology, for example in the work of George Katona, one of the forefathers of modern economic psychology. Katona focused on the ways in which many of our personal goals interact with group goals. The power of the group is determined by how powerfully each member identifies with the group. Katona theorised that this will determine how individuals interact with groups during herding and social learning. There will be feedbacks between the individual and the group. As individual group members imitate each other, this reinforces the coherence of the whole group.[22] Football fans are an everyday example of this phenomenon. When they emulate their team and other fans – by buying and wearing the same football strip, for example – this reinforces the cohesiveness of the entire football club. The football club needs its fans as much as the fans need their football club.

Mob identities

If collective herding is not driven by our self-interest, why is it so powerful and cohesive? Identity is one of the essential factors determining the power of groups and collective herds. Theories of identity are captured in different ways across the social sciences, and we can rethink identity in the light of evidence derived from psychology and sociology alongside economic analyses.

Consistent with some of Le Bon's insights, we can understand identity as something more diffuse than the economic concept allows. As we saw in chapter 1, economists such as Akerlof and Kranton have developed an economic approach to identity, describing it in transactional terms – as a form of social exchange.[23] A rational, self-interested individual signals to a group to demonstrate that they share an identity with that group, and they do this in order to benefit from the support the group can give them. So, in economics, identity is essentially determined by each person's cost-benefit calculations about what they will gain from joining a group. In social psychology, identity is not obviously concentrated around the net benefits to self-interested individuals. The identities that bind groups together are not so much about advantages for each individual. Rather, they relate to how the group as a whole can build strength through its sense of identity. Identity determines how we interact with different groups around us. In the language of social psychology, we identify with our *in-groups*, and so we tend to favour them. We do not identify with our *out-groups* and so we are inclined to discriminate against them. We feel a strong social bond with our in-group, and we will copy and emulate its members, even with practices such as tattoos and other forms of painful body modification that seem perverse to outsiders. The decisions we make to build a sense of identity with specific groups may take other, more benign forms – perhaps wearing certain types of clothes or buying specific types of consumer goods. Our consumption choices are not just about increasing our own satisfaction, they are also about building a sense of shared identity, and encouraging members of a group to act in concert. With a strong sense of identity, the group is more likely to be strong and robust.

What drives us to build our identities around one group rather than another? How do we decide who is in our in-group and who is in our out-groups? The Polish social psychologist Henri Tajfel tackled these questions. Like Wilhelm Reich, Tajfel's work was inspired by the destructive power of fascism. As a Jew, he was excluded from the Polish university system and so studied in France, and served in the French army during the Second World War. He was captured by the Germans and spent some time in prisoner-of-war camps, returning home after the war to discover that his entire family and most of his friends had been killed in the Holocaust. This inspired him to think deeply about how racism, prejudice and discrimination have their roots in questions of identity. He developed his *social identity theory* in an attempt to understand the persecution of Jews, not only by the Nazis, but also by non-Nazi mobs of ordinary Germans.[24] How was the majority of the German population so easily persuaded to comply with the diktats of Hitler and the Nazi Party? Why did so many of these ordinary people identify so easily with such an extraordinarily vicious cause?

Tajfel's research focused on two aspects of identity. First, he found that we form bonds with others very quickly and easily: it does not take very much at all to encourage us to identify with some groups and not others, even when we have a minimal amount in common with them. This underlies Tajfel's *minimal group paradigm*, and it helps to explain how mob psychology forms rapidly and unconsciously.[25] Essentially minor choices can operate as a surprisingly powerful signal in building allegiances with our in-groups. With minimal encouragement, we join a group with which we identify, even if that group is formed around mendacious principles.

Second, we tend to discriminate against our out-groups. Affinities with some groups and antipathies against others fuel tensions between in-groups and out-groups. We can be encouraged to copy those around us engaging in discriminatory, prejudiced behaviour against other groups.

Tajfel and his research team developed a series of pathbreaking experiments, exploring how easy it is to polarise people by building divides on fragile grounds.[26] They brought together a group of sixty-four boys who knew each other well, from the same house and form at a comprehensive school in Bristol. The boys were split into groups of eight. In the first stage of the experiment each boy was shown forty clusters of dots, each containing a varying number of dots. The boys were asked to estimate the number of dots in each cluster. They were given some spurious information about the motivation behind this first experiment. Then the boys were told that they would be participating in another, unrelated experiment, which was really designed to capture how easily the boys identified with a specific group. They were told that, for convenience, they would be assigned to one of two groups according to the similarity of the answers they had given to the dots task – whether they had underestimated or overestimated the number of dots. In reality, the boys were randomly allocated to the two different groups. On this minimal and false basis – the random allocation of boys into one group of 'underestimators' and another group of 'overestimators' – the boys formed their in-groups and out-groups.

Now that the boys' bonds were formed, the next stage of the experiment tested their favouritism in discriminating against out-groups. The boys were asked to allocate financial rewards and penalties to other boys. They were not told the identities of those they were rewarding or penalising, only that they were giving or taking from either an underestimator or an overestimator. No boy had anything to gain for himself so the allocations could not be determined by each boy's self-interest. It was found that the boys consistently favoured those in their in-group. They gave the other boys in their in-group a bigger share of the money than they gave the boys in the out-group.

Tajfel and his team then tested the robustness of this finding by separating the boys according to their preferences for 'foreign painters' – either a painting by Paul Klee or one by Wassily Kandinsky (though the boys were not told the names of the painters). Their findings were broadly the same as those from the dots experiment. When the boys had a choice between maximum profit for all the boys together and maximum profit for their in-group, the boys tended to favour their in-group. Tajfel's experiments showed how easy it is to engender a sense of loyalty to a group, even on the seemingly spurious basis of patterns of dots or artistic preference. There is no obvious reason, at least from the perspective of self-interest, to form affinities on such flimsy bases. This underscores the insights from Le Bon and Reich that mobs cannot be easily understood if we do not take the perspective of the group as a whole in itself. Collective herding cannot be understood just by looking at individual herd-members because individuals in herds are prepared to discriminate against out-groups even when it is not in their individualistic self-interest to do so. With collective herding, group goals are paramount.[27]

Do Tajfel's findings apply in more ordinary contexts – such as choosing to grow a beard or style our hair in a topknot? That these superficial choices work so well in building identity is consistent with Tajfel's minimal group paradigm. Modern hipsters are a salient example.[28] Outwardly unconventional, hipsters are rebelling against out-groups, but conforming and identifying strongly with a specific in-group. They are simply copying a small in-group by dressing the same way and conforming to (minority) conventions. With Tajfel's evidence in hand, we can return to the economists' conceptions of identity introduced in the previous chapter. We saw there that economists focus on costly signals – for example painful permanent face tattoos. From an economics perspective, why would an individual incur large costs (physical, economic and/or monetary) if there was nothing in it for them? In economic models of identity, in-groups will believe that an individual who has incurred significant psychic as well as economic costs in getting face tattoos is sincere about their membership of the group. Costly signals make economic sense because they are more credible. Tajfel's findings undermine this explanation, however. Tajfel and his colleagues showed that group identity can be formed without people having to do very much at all to signal to the groups with which they identify. Whilst hipsters and other rebels want to be defined as different, they do not need to incur significant personal costs to persuade others that they belong. More broadly, identity does not need to be founded on demonstrably strong political and ethical convictions. People can disingenuously adopt the uniforms of rebels and outsiders with much the same impact as if they had got themselves a face tattoo. As journalist and blogger Ezra Klein argues, replying to economist Paul Krugman's observations about some hipsters at a music festival,

> Krugman suggests that hipsters are signaling a rejection of the workaday bourgeois world by flouting conventional dress codes. I think the truth is closer to the opposite: They're signalling a mastery of the workaday bourgeois world by flouting conventional dress codes . . . as venture capitalist Peter Thiel writes . . . 'Never invest in a tech CEO that wears a suit.'[29]

In essentially superficial ways, hipsters can cheaply signal to their potential investors that they are creative nonconformists. Outside the business world, when we join groups of other copycats it is not necessarily difficult to signal to them that we belong. We can join a herd without incurring any immediate costs, whether tangible or intangible. The collective herd can build momentum without any individual member having to consider carefully what they're joining and why.

Mobs at night-time

The impacts of identity on groups and mob psychology are themes of enduring interest, not only for academics but for social policymakers too. Our night-time lives are often associated with potentially violent and antisocial group behaviours. Exeter psychologist Mark Levine is interested in how people interact in the night-time economy, especially after the pubs close. In the past, policymakers have assumed that violence escalates when pubs all close at the same time. In the UK in 2003, for example, Tony Blair's government relaxed some of the licensing laws to allow for staggered closing times, to reduce the size of unruly drunk crowds. If lots of drunken revellers simultaneously spill into city streets, they thought, then the chance of violent altercations is magnified.

Levine and his colleagues focused on the idea that late-night mob violence is not usually about individuals fighting with other individuals. It is about conflicts between in-groups and out-groups. Aggressors are often motivated by a desire to either show off to their in-group or threaten their out-groups, and their impulsive violent actions are fuelled by alcohol.[30] In other words, this is a demonstration of collective herds in conflict, with each possessing an identity that is not easily explicable in terms of its individual members but which plays an important role in the conflict itself. And this is collective rather than self-interested herding because late-night violence does not obviously relate to what any lone person can gain in terms of individual self-interest.[31]

To test their hypotheses, Levine and his colleagues concentrated their study on three cities in the northwest of England. They convened twenty focus-group interviews with fifty-three people – a mixture of students, manual and retail workers, and a handful of people on prison probation. From these group interviews they collected seventy-seven stories in which the interviewees recounted their direct experiences of violence, either as participants or observers. Only direct witnesses were included because the researchers wanted to know about first-hand experiences. Their findings were both expected and surprising. Stranger-on-stranger violence was rare. Fights between members of the same in-group were common. One interviewee commented, '[Y]ou can know people too well . . . because he's your mate and because you're drunk, then you get aggressive with [him].'[32] Intra-group conflicts were interpreted as friendly banter. They were often quickly resolved and forgotten.

Violence was more serious when people fought with out-groups. Intergroup violence was usually driven by mob psychology and the group's interests, and not by individuals' independent actions. One interviewee observed that 'Instead of single people, it's gangs of lads.'[33] Another interesting and perhaps unexpected finding was that many of the interviewees did not believe that police intervention was necessary most of the time. The collective herds were, to an extent, self-regulating, reflecting people's strong social instincts to help others in distress. Often people observing a fight would play a positive and effective role in intervening and de-escalating the violence. Levine and his colleagues also noted that people watched out for their friends during nights out, and monitored whether they were drinking excessively. So, whilst mobs and crowds might play some role in escalating violence, they are also able to monitor and regulate themselves. Levine and his colleagues concluded that crowds may have positive as well as negative impacts. The presence of collective herds in the night-time economy is not unequivocally bad. In fact, perhaps police intervention during the fights simply magnifies the opportunities for conflict and violence, by introducing another out-group into the fracas.

Peer pressures

We have seen that mob psychology reflects interplays and feedbacks between individuals and groups. Identity plays a crucial role and the extent to which we identify with our in-groups against our out-groups helps to explain why tensions between different groups emerge. We can form bonds with our in-groups very easily, but are there other psychological explanations for our tendencies to conform so easily? What encourages an individual to do what others are doing, even when their choices do not align with their ethical principles or own self-interest? Groups need to develop ways of reinforcing group norms – behaviours that prioritise group interests over individual interests. Peer pressure plays a powerful role in this, helping to ensure the cohesiveness of groups, crowds and mobs.

The social psychologist Solomon Asch conducted a range of pioneering and influential experiments to demonstrate the power that peer pressure exerts in group settings. Like Tajfel, Asch was a Polish social psychologist of Jewish origin, but his family left Europe before the Holocaust, immigrating to New York in the 1920s. Asch completed his high school and college studies there and went on to have a distinguished career as an academic social psychologist in the US. In the 1930s, on hearing of Hitler's hypnotic influence over the German population, he hypothesised that Nazi propaganda was effective because it tapped into an unconscious combination of fear and ignorance. He went on to develop an interest in our susceptibility to social influences, particularly when we are processing new information, including propaganda. To colleagues, Asch also recounted another event from his childhood that had fuelled his interest in conformity. One Passover night, he had been allowed to stay up late. He watched as his grandmother poured an extra glass of wine, and Asch's uncle explained that this last glass was for the prophet Elijah. As an impressionable child, Asch thought he saw some wine disappear from this extra glass. At some unconscious level, he was responding to group pressure from his family, forming a superstitious belief in the prophet's intervention. So he thought that Elijah really had taken a sip, his imagination fuelled by his instinct to conform to his family's beliefs.[34]

Asch and his team designed a line judgement task to test for peer pressure.[35] They wanted to discover if people could be manipulated into giving obviously wrong answers to simple questions just because they felt a real or imagined peer pressure from a group around them. Asch's experiment has since been replicated and adapted extensively, but the initial experiment was simple. Groups of between seven and nine male college students were assembled in a classroom and shown a series of two cards – we'll call them Card A and Card B. Card A depicted a single line. Card B depicted three lines of different lengths. The students were asked to choose which of the three lines on Card B matched the line on Card A. They had to announce their answers to the rest of the group, one by one. The experiment was then repeated numbers of times.[36]

In the early rounds, everyone got it right (it is a very simple task, after all). In the third round, however, the scenario changed. One of the students was surprised to find that he disagreed with the others in his group about which line from Card B matched the line from Card A. He did not know that the experimenters had briefed the other students to give the same wrong answer. In each group, the lone student confronted a conflict between his own beliefs and the unanimous judgement of everyone else.

Asch and his team conducted this initial experiment across three academic institutions, with 123 students placed in the minority scenario outlined above, and the experimenters talked to the students afterwards to find out more about how they had reacted to the confusion. The lone minority students changed their answer to match those of the lying majority 37 per cent of the time. Individual differ-

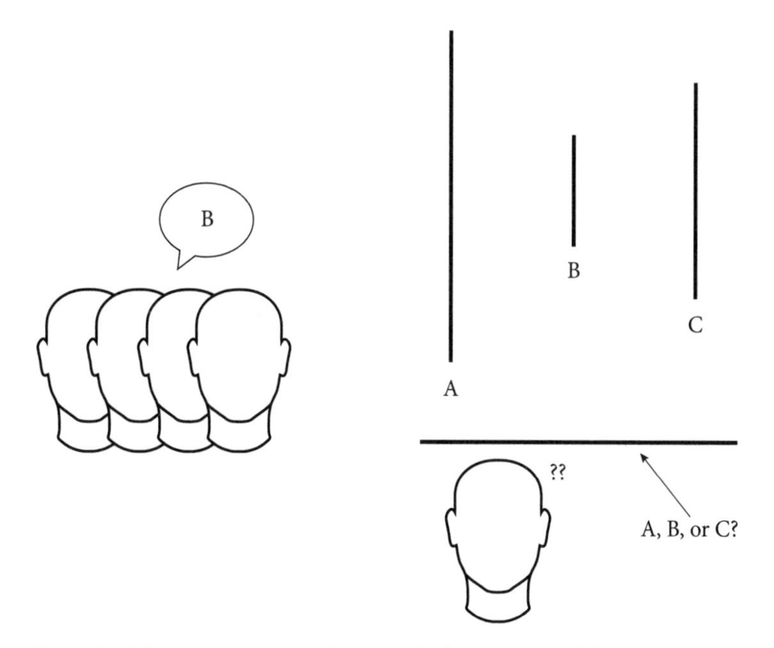

Figure 3. A line experiment: a subject is asked to guess which line matches the horizontal line, when the herd says 'B'.

ences modulated the students' responses, suggesting that personality and emotions play key roles in determining whether we decide confidently or otherwise to be copycats or contrarians. Asch and his team loosely separated their student participants into categories according to their emotional responses. Some students were admirably independent and did not seem to worry about being in a minority. They did not respond strongly to the majority opinion and seemed easily able to adapt quickly to the doubts raised by others, calmly retaining confidence and sticking to their own initial (and correct) answer. Other students expressed significant distress and confusion when they found that they were in a minority of one. One group of 'dissenters' did not sway from their correct answers, but being in the minority worried them. They became confused and unsure and stuck reluctantly to their correct answer. Finally, there was a broad category of students described by Asch as 'extremely yielding persons', who were persuaded by the group's response to give the wrong answer. In their post-experiment interviews, these yielding students rationalised the disagreement in different ways. Some blamed their mistakes on the other students, arguing that the others' sheep-like behaviour had been misleading. Some students thought that perhaps the experimenters were trying to trick them with an optical illusion. A further self-critical group thought that their initial answers had been the product of their own stupidity. Asch and his team also noticed that the students who yielded to the majority answer systematically underestimated how often they were conforming to the wrong majority answer, perhaps suggesting unconscious influences were at play.

For researchers, interpreting the findings from Asch's experiments is not easy. The conforming behaviour that Asch and his colleagues observed could be attributed to one of two types of social influence. To recap from the introduction: informational influences are about following others because we believe that others' actions are informative; and normative influences are about us feeling a less concrete and more unconscious need to conform to peer pressures and social norms. Failing to conform generates awkwardness and can lead to confrontation and confusion. Conformity is much easier: it provides psychological reassurance, and is psychologically satisfying, especially if it means we can minimise inter-personal conflicts.

So, are the participants in Asch's line judgement experiments responding to informational influences or normative influences? Are they worrying about what others will think of them, and agreeing with the group because of social norms and sociopsychological factors more generally? Or are the line-judgement task participants in fact trying to learn something by observing others' behaviour, consistent with the models of self-interested herding from the previous chapter?

The Nobel Prize-winning economist Robert Shiller has argued that following others in giving wrong answers in simple tasks is not inconsistent with rational social learning.[37] People may, rationally, discount the accuracy of their own judgements if they see a lot of other people coming up with a different answer. One possible explanation is that the students were using a Bayesian reasoning process to balance different bits of information, as described in chapter 1. Self-interested herders, engaging their social learning faculties, could rationally conclude that there is only a slim chance that they are right and everyone else is wrong. Shiller quotes one of Asch's participants explaining, 'To me it seems I'm right, but my reason tells me I'm wrong, because I doubt that so many people could be wrong and I alone right.'[38] Particularly in situations of uncertainty, when people have little faith in their own judgements, they will overestimate the accuracy of other people's.

Shiller also notes that findings similar to Asch's have been identified in studies of human–computer interactions. If participants behave in similar ways outside a human-to-human context, then perhaps this suggests that personal social pressure was not the key influence and participants were using logic and reason to balance their own judgement against those of others. But what if people engage with computers as if they are real people? Then Shiller's justification heads into the territory of unfalsifiable hypotheses. We could use a similar logic to justify any action as rational, without having empirical evidence to verify it. We cannot objectively refute a psychological explanation grounded in unconscious sociopsychological motivations based on humans' interactions with computers. Whilst it may be hard to imagine what sort of experiment could be designed to separate completely the economic and psychological explanations, neuroscience is giving us deeper insights into these and other types of decision-making conflicts. In the next chapter, we shall explore how neuroscientific tools such as brain imaging can be used to unravel these conundrums, giving us more and richer information about whether copycats and contrarians are driven by instincts and emotions, cognition and deliberation, or some combination of the two.

Learning social norms

Another form of sociopsychological influence comes from social norms, which differ from peer pressure because they are more diffuse and enduring. Social norms are sticky – in other words, they are hard to shift. This allows them to operate even when we are not directly in contact with the group. If social norms operate even outside group settings, where do they come from? They operate at a deep unconscious level, sometimes reflecting influences from our childhoods. Children's behaviour often mirrors that of the adults around them as they learn by observing others. This observational learning is driven by our ingrained instincts to imitate. Psychologist Albert Bandura explored these ideas in constructing his *social learning theory*. Bandura focused on the role of cognition in imitation, particularly amongst children. He identified a link between the aggressive behaviour of children who had earlier observed aggressive behaviour in adults. In his experiment, Bandura and his team left groups of toddlers to play in a room full of toys and exposed them to three different scenarios. In the first 'aggressive' scenario, the children played while an adult in the room behaved aggressively towards a doll. In the second 'non-aggressive' scenario, an adult in the room was playing quietly and non-aggressively. In the third 'control' scenario, no adult was present. Bandura and his team discovered that the children in the aggressive scenario, who had had an opportunity to observe an adult's aggression, were more likely to imitate the adult's violent behaviour in their own play. The children's acts of aggression mimicked the specific physical actions of the adult, suggesting that children's instinct to imitate adults has a strong influence on their behaviour.[39]

Social pressure as a policy lever

We have seen a range of ways in which economic incentives and psychological influences can feed into our instincts to imitate and form groups, herds and crowds. But, so what? Why are these insights useful? They are useful because people's susceptibility to peer pressure can be used as a policy tool, to moderate the negative impacts that some of our behaviours have on communities more widely. Whether learning by imitating others or deciding collectively, herding sometimes enables better decision-making, from the perspective of both the individual and the group. We are social animals and are generally rewarded for behaving in a prosocial way, so social norms have a powerful influence on our behaviour. If teenagers copy their peers in their choices and habits, then they are probably more likely to be invited to the coolest parties. From an individualistic perspective, sometimes our own self-interest will be promoted if we conform to the norms of the herd. Social norms are built around others' behaviour because other people around us give us our standards for behaviour. We compare our own behaviour with what others are doing, and others' behaviour provides us with what behavioural economists call our *social reference points*. We make our own decisions by reference to what we believe to be the average, conventional decision of the group. We do this either because we believe that larger numbers of people agreeing with each other are more likely to be correct, and/or because belonging to a group strengthens our sense of belonging.

Many organisations, from marketers to government policymakers, use peer pressure and social reference points to leverage copycats' conformist natures. A range of research studies, including a large-scale study of OPower customers in California, showed that many (though not all) people are likely to reduce their energy consumption if they think their consumption exceeds the average of their friends and neighbours.[40] In the UK, Her Majesty's Revenue and Customs found that taxpayers were more likely to pay a late bill if they were told that they were in a small minority of late payers. Information about crowd behaviour often encouraged taxpayer conformity, though not always.[41]

Our conformist instincts have also been harnessed for public health improvements in low income countries. Sanitation habits are an essential ingredient for public health: disease is reduced when people defecate into latrines rather than in open public spaces. The World Health Organization (WHO) has explored the impact of peers' opinions on people's existing habits as a tool to improve sanitation – especially in underdeveloped rural regions where there are high levels of infant mortality.[42] A WHO team of social researchers recognised that economic incentives and disincentives, such as subsidies and fines, have little impact on sanitation habits when social norms and traditions are strong. They also suspected that it is not enough for people just to know things to change their behaviour. To investigate these ideas, the researchers designed and implemented a field experiment in Orissa, India. They targeted 1,050 households across 20 villages, rolling out an information campaign educating people about the importance of sanitation, clean water and good hygiene. To test the idea that knowledge is not enough to change ingrained behaviour, the researchers included a treatment condition in their experiment. They combined their education campaign with a deliberate attempt to tap into people's unconscious instincts via a social trigger to leverage people's social emotions. Thus, the WHO's 'Community Led Sanitation' scheme incorporated a 'walk of shame' during which all members of the community would walk together and identify instances of poor hygiene along the way. The team also developed 'defecation maps', with the villagers helping to identify the spatial distribution of defecation. The volume of faecal matter was calculated and discussed amongst the villagers, along with information about its likely impacts.

The WHO's 'shame or subsidy' policy tapped into psychological influences to encourage the use of public sanitation infrastructure, funded via development initiatives from international multilateral organisations including the World Bank. The policy was effective but controversial. In some villages, latrine use increased from 6 per cent to around 30 per cent. Public shaming triggered social emotions, and peer pressure worked to change people's ingrained habits – habits that were harmful to them and others around them. This evidence was used by the WHO to advocate policies for improving people's sanitation habits based around 'social marketing' – a euphemism for using social pressure and peer monitoring as policy tools. But the ethical dimensions of this study and the consequent policy implications are complex. Was it appropriate for policymakers to manipulate behaviour by using people's relationships with each other – however well intentioned? Whatever the answer, the WHO evidence does show that our copycat natures and our susceptibility to peer pressure can be an effective complement to traditional economic policy instruments, including taxes and subsidies, in improving people's living conditions. The power of these solutions is not about appealing to our self-interest. It is about tapping into our unconscious sociopsychological drivers, including our susceptibility to the influence of others around us.

In this chapter we have explored the many ways in which mob psychology distorts our behaviour. We have also explored how this links with the concept of collective herding, in which group behaviour is not explicable in terms of the individual self-interest of the herd's members. Insights from psychology help us to understand why and how collective herds seem to have minds and missions of their own, and why individuals lose their sense of self when they join a collective herd.

When we copy others, are we just being logical and self-interested? Or are we driven by some unconscious psychological instinct to imitate and conform? Considering the different explanations for self-interested herding versus collective herding, as outlined in this and the previous chapter, what can we conclude about the relative power of economic and sociopsychological explanations? Do other social sciences capture these group behaviours more powerfully than economics? Yes and no. In contrast to the economists, psychologists and sociologists focus much more on how and why personality, emotions and social norms drive our choices to join herds, mobs and crowds. They can explain collective herding. They also explore a range of other more diffuse and unconscious forces. These influences are powerful, not only during extreme episodes of collective madness such as the Jonestown massacre, but also in more ordinary situations in which we choose to lose our personal autonomy and ignore our own self-interest by joining a group. But whilst peer pressure, identity and group influences are crucial in understanding mob psychology, we should not forget the economists' models of self-interested herding. In many contexts, we have more straightforward and logical motivations and incentives to follow others. Economic goals and incentives are important motivators too.

In the next two chapters, we will introduce some studies from the behavioural and biological sciences, including cognitive neuroscience, evolutionary biology and behavioural ecology. Scientists working in these fields have added new and fascinating dimensions to our understanding of copycats and contrarians. They have also shown us ways to combine the divergent explanations from economics and the other social sciences. With the broader understanding of human motivations and drivers enabled by a more general theory, we should be able to smooth away some of the apparent contradictions between the economists' conventional models of self-interested herding and other social scientists' models of collective herding.

3

Herding on the brain

In an allegory written in 360 BC, Plato imagines a dialogue between Socrates and an Athenian nobleman named Phaedrus. The pair sit together under a plane tree on the banks of the Ilissus river in Athens. Socrates contemplates madness. He explains to Phaedrus the nature of the soul, in both its human and divine forms. Socrates postulates that the human soul is a chariot – a pair of winged horses driven by a charioteer. The first horse is 'noble' and 'good', the second 'ignoble' and 'bad'. And our charioteers struggle to control the ignoble horse:

> The right-hand horse is upright and cleanly made . . . he is a lover of honour and modesty and temperance, and the follower of true glory; he needs no touch of the whip, but is guided by word and admonition only. The other is a crooked lumbering animal, put together anyhow . . . [he] is the mate of insolence and pride, shag-eared and deaf, hardly yielding to whip and spur . . . heedless of the pricks and of the blows of the whip, [he] plunges and runs away, giving all manner of trouble to his

> companion and the charioteer . . . he persists in plaguing them, they yield and agree to do as he bids them . . . [The horses are] carried round below the surface, plunging, treading on one another, each striving to be first; and there is confusion and perspiration and the extremity of effort; and many of them are lamed or have their wings broken through the ill-driving of the charioteers; and all of them after a fruitless toil, not having attained to the mysteries of true being, go away, and feed upon opinion.[1]

What has this got to do with copycats and contrarians? The divergent accounts of self-interested herding and collective herding seem as irreconcilable and mutually exclusive as the noble and ignoble horses, and we are left in a quandary. Are the economists, focusing on reason and deliberation, correct to assume that herding is a rational, individualistic choice formed by our capacity for logical reasoning? Or are psychologists and sociologists, focusing on collective herding as the outcome of ephemeral emotions and socio-psychological instincts, correct to emphasise what some would call the 'irrational' aspects of our behaviours? Plato's allegory is interesting because it suggests that both approaches have merits. If we can bring them together then, potentially, we will have a much more powerful account of herding. We might be able to develop a more general theory to capture the rich and myriad ways in which our copycat and contrarian natures interact in our daily lives.

A key problem for social scientists studying social behaviour and crowd psychology is that we have not been able to see how copycats and contrarians reach their decisions. We can observe what people choose, but without knowing the deeper processes underlying these decisions and actions. For economists specifically, the human brain has been like a black box.[2] We may know what people know and we can observe their choices but we cannot see how the brain processes the information before a person's choices are revealed. For this reason, empirical economics has tended to focus on quantifying people's observed behaviour (a preoccupation it shares with behavioural psychology). Evidence about people's actions is objective. It can be counted, collated and stored in statistical agencies' databases. More recently, experimental evidence from ordinary lab experiments has been added to the stores of data, but a lot of experimental evidence is also, essentially, about observing what people choose to do, and fails to capture the underlying psychological mechanisms. For a long time this was as much as social scientists could hope to do while people's thinking processes were largely unobservable.

With modern science, however, these constraints are unravelling. The biological sciences can help to fill the gaps in our understanding of our drives and motivations by illuminating how we think about our decisions and choices. Neuroscientists have developed some interesting theories and tools that illuminate how different thinking styles interact when we join crowds and herds. They can show that different parts of our brains are activated in different contexts. We engage different brain areas when we are feeling emotions, and these brain areas are distinct from, but sometimes complementary to, the parts of our brain that are activated when we are thinking analytically. Reflecting Plato's early speculation about the different facets of our souls, reason and emotion do not operate independently. Capturing the complex interactions between them not only adds to our understanding of copycats and contrarians. It also illustrates that social scientists' debates about whether herding is driven by rational or irrational influences are increasingly redundant.

Personality struggles

Plato's suggestion that opposing forces within our personalities are driving us has been a theme throughout intellectual history. Some of our modern thinking about personality struggles has its origins in Sigmund Freud's work, which we introduced in the previous chapter, though modern scientists strive to be more objective and empirical. The idea that our choices are driven by an interaction of different thinking systems is now re-emerging alongside empirical tools to test the power of these hypotheses. Economic psychologist and economics Nobel laureate Daniel Kahneman has spent his career exploring psychological influences on decision-making, and popularised the key insights in his 2011 book *Thinking, Fast and Slow*.[3] Kahneman distinguishes between reason and emotion but, as is the norm in modern science, he crafts his analysis of the duality of our character in less judgemental language than Plato's. Competition between our different thinking styles is not about a battle between good and evil, between our noble and our ignoble souls. Sometimes reason is a good guide, sometimes emotion is a good guide, sometimes the best guide is a combination of reason and emotion together.

Kahneman sets out his dual systems model in terms of interactions between two different thinking styles: System 1 and System 2. System 1 thinking is quick, automatic, intuitive and emotional. When we come across a wild lion in the bush, System 1 is in the driver's seat. We feel fear, and we run or hide without consciously considering our options. System 2 thinking is quite different. It is slow, controlled and deliberative. In situations when cognitive effort is vital, then System 2 thinking will step up. When we are in a job interview, sitting an exam or playing chess, then System 2 is in control, and we draw on our logical, reasoning capacities.

System 1 thinking requires much less mental energy than System 2. Conversely, System 2 is good at deliberation and carefully assessing different options, but it is lazy and wants to economise on cognitive effort. As Kahneman observed:

> most of what you ... think and do originates in your System 1, but System 2 takes over when things get difficult ... The division of labor between System 1 and System 2 is highly efficient: it minimizes effort and optimizes performance.[4]

So Systems 1 and 2 do not operate alone. They act in concert, but the quicker System 1 dominates most of the time. Reason is not irrelevant when we are in danger. Emotion is not irrelevant when we are forced to think deeply. Both will be operating, either in the foreground or the background of our thinking.

Kahneman's analysis of different thinking styles is useful in our study of copycats and contrarians. It can be applied to capture how our herding and anti-herding choices are motivated by interactions between System 1 and System 2 thinking, connecting the self-interested herding models of the economists with the collective herding models from other social sciences. As we have seen, self-interested herding is about inferring something about what motivated others around us to make their choices. We balance this social information with what we know (our private information) and use logical rules (such as Bayes' rule) to reconcile discrepancies between private and social information. All this is led by a System 2 style of deliberative thinking. Collective herding is driven by deeper, less conscious influences including emotions, personality, psychological instincts and social pressures. With collective herding, System 1 is in control. Which system dominates will depend on the situations in which we find ourselves. When we need to decide quickly, collective herding is more likely to dominate. When we have more time to reflect, self-interested herding will dominate. Sometimes the two will be operating together, as we shall see from the neuroscientific evidence. Similarly, anti-herding contrarians also sometimes deliberate slowly and carefully, but at other times decide to rebel, triggered by impulsive, instinctive emotional drivers.

Measuring mimicry

We have explained how dual systems models can reconcile divergent explanations for herding.[5] These theories have a lot of power, but they do raise some empirical questions – paralleling those we might ask about Freud's attribution of our adult behaviour to unconscious drives formed from our childhood experiences. Just as it is difficult to empirically verify a Freudian account, how can we provide evidence about whether System 1 or System 2 is in control? How can we *know* whether herding and anti-herding reflects careful deliberation, or emotional impulse, or some combination of the two? We cannot necessarily tell which system is driving someone just by observing what they do.

To answer these questions, we need to know more about neuroanatomy and its links with the basic principles underlying modern neuroscientific techniques. Today's understanding of neuroanatomy builds on neurological insights from Aelius Galenus, better known to us as Galen (AD 129–*c.* 199), an impressively insightful physician whose work foreshadows many insights from modern neurology. Galen was born in Pergamon (an ancient Greek city, now part of modern Turkey) into an affluent family. His architect father, Aelius Nicon, had initially pushed his son towards philosophy and politics, but had a dream in which Asclepius, the Greek god of medicine, instructed him to allow his son to study medicine.[6] Galen went on to develop a successful medical practice in Rome. He was physician to Marcus Aurelius' son Commodus and became part of Rome's intellectual community under a succession of emperors.

Galen's knowledge of neuroanatomy was enhanced through his work as a surgeon, including a spell tending to the gladiators of Pergamon. Galen was influenced by Plato and developed the Greek philosopher's chariot allegory in ways that are pertinent to the idea that our thinking styles might be rooted in our brain structures. Foreshadowing Freud's id, ego and superego, Galen thought that our brains are the home of rational thought, our spirituality is in our hearts, and our appetites are in the liver. His medical practice complemented his interests in how our brains work – remarkably, very early on he recognised that the spinal cord is an extension of the brain.[7]

Many centuries later, Gustave Le Bon, whom we met in the previous chapter, developed insights that were similiar to Galen's. For instance, he postulated that the spinal cord channels the social emotions manifested in mobs whilst the brain guides orderly and rational crowd behaviours.[8] Some of Le Bon's speculations around neuroanatomy – his theories to do with brain size and intellect across genders and races, for example – are discredited in modern neuroscience.[9] Nonetheless, Galen was on the right track with his ideas about how brain structure links to the psychology of crowds and mobs. Galen's and Le Bon's hypotheses would strike many modern neuroscientists as gross oversimplifications, especially Le Bon's very rough division of the spinal cord from the brain. He did, however, anticipate some findings from modern neuroscience. Neuroscientists have now identified regions deep in our brain associated with more primitive and emotional thinking, linking areas in our brain stem and midbrain limbic system with our impulsive and/or social behaviours. Areas in our prefrontal cortex (the region at the front of our brain, above our eyes) have been implicated in tasks that require more complex thinking, including mathematical and analytical reasoning, and economic decision-making.

Opening black-box brains

The tools that neuroscientists can use to unravel what is going on in the black boxes of our brains are increasing all the time in range and sophistication. How can they capture the underlying neural processes that drive our choices, including our tendencies towards herding and anti-herding? Some of the early applications of neuroscientific tools were based around lesion patient studies. These studies focus on people who, through either accident or illness, have experienced localised brain damage. Using information about the location of the damage, neuroscientists can make inferences about how those brain areas are implicated in different types of decision-making.

Galen himself conducted some very early lesion patient studies, having been puzzled by the fact that no-one had 'ever taken the trouble ... to put a ligature around parts of the living animal in order to learn which function is injured'.[10] Galen's experiments did not go much further, however, as he came up against both religious and scientific constraints. Lesion patient studies resurfaced after the Enlightenment as science started gaining ground over religion. A famous historical lesion patient was Phineas Gage, an American railway worker, who in 1848 suffered a harrowing accident. A tamping iron, used to pack explosives into holes, exploded and was shunted into the front of his skull and through his brain. Amazingly, Gage seemed to recover well from his accident – at least physically. However, his friends and colleagues started to notice significant changes in his personality. A reliable and industrious worker, Gage had held down a steady job for years, but after recovering from the accident he was not such a good employee. His personality had changed. At work, he became feckless and unreliable. Socially, he became erratic and difficult. His physician Dr John Martyn Harlow was fascinated by these changes in Gage's personality. He studied Gage and his medical record intensively and concluded that the change in his patient's behaviour could be explained by the damage sustained to the frontal lobes, the areas of our brains associated with higher levels of cognitive functioning and self-control.[11]

More than 150 years later, modern neuroscientists are drawing on similar studies extensively. The US-based neuroscientist Antonio Damasio and his colleagues are pioneers in the use of lesion patient studies to study economic and financial choices. They are especially interested in what guides our risky choices, for example in gambling or asset trading. Damasio and his team have presented much evidence about the important role that emotion plays in decision-making, demonstrating that brain lesions in emotional processing areas are associated with severe deteriorations in ordinary functioning, even for patients with no outward evidence of injury. Mirroring Kahneman's model of dual systems thinking, Damasio argues that emotional influences do not necessarily preclude rational thought.[12]

Lesion patient studies are relatively simple, if blunt, tools. Neuroscientists cannot directly control the regions available for study (unless they are complicit in significant legal and ethical transgressions, forbidden by modern research ethics committees). Unfortunate accidents and illnesses dictate which areas of the brain are damaged and neuroscientists are confined to studying the lesions as they find them. In the last few decades, however, the technological sophistication of the neuroscientist's toolbox has rapidly advanced. Improvements to physiological and neuroscientific techniques mean that we can start to observe and understand how our neural circuitry is responding as we make our decisions. Physiologists can monitor heart rate, skin conductance, sweat rate and other physical responses and use this evidence to make inferences about emotional responses. Neuroscientists can measure brain activity by using techniques such as electroencephalography (EEG) to capture electrical impulses on the scalp. They can measure blood flow through the brain using brain-imaging techniques. They can zap areas of the brain temporarily to disable them using a technique called transcranial magnetic stimulation.

Brain imaging is a particularly popular technique. It requires complex machinery, but it gives neuroscientists more control over which areas of the brain they can study. Brain scanning also enables neuroscientists to work with a broader range of healthy people, thus addressing the ethical concerns around experimenting with vulnerable patients. Brain scanning techniques are used to capture how blood flows into localised regions of the brain. When we respond to mental stimuli, specific brain regions are activated, and blood flow in these areas increases relative to blood flows through passive brain regions. This produces changes in magnetic susceptibility, which can be mapped using a magnetic resonance scanner. This scanning technique is known either as Blood Oxygen Level Dependent (BOLD) brain imaging or functional magnetic resonance imaging (fMRI). Brain scanning is far from infallible and is often prohibitively expensive.[13] It does, however, allow neuroscientists to focus on what is happening in specific brain areas. With fMRI, neuroscientists can study brain function in a targeted and controlled way, including as people participate in specific activities and tasks. By identifying specific 'regions of interest' in the brain, and by separating out the areas usually implicated in emotional, instinctive decision-making from those associated with higher-level cognitive reasoning, fMRI studies can capture whether herding is driven more by our emotional System 1 thinking or our deliberative System 2 thinking, or some combination of the two.

Copycats and contrarians in the brain scanner

In applying some of these techniques to discover more about the thought processes driving copycats and contrarians, we can learn some lessons from other brain imaging studies. One pioneering fMRI study of System 1 versus System 2 thinking was conducted by Dutch neuroscientists Wim De Neys, Oshin Vartanian and Vinod Goel.[14] They used imaging techniques to investigate some judgement tasks that Daniel Kahneman and his old friend and colleague Amos Tversky had devised in their early work, specifically to see if these connected with Kahneman's more recent ideas about dual thinking systems. De Neys and his colleagues used a version of Kahneman and Tversky's Engineer-Lawyer problem.[15] Participants in this experiment were told that a sample of 1,000 people includes 5 engineers and 995 lawyers. The probability that a given person is an engineer is 5 in 1,000; the probability that they are a lawyer is 995 in 1,000. The participants were then asked to estimate the chances that one person from this sample is either a lawyer or an engineer. Alongside the statistical information, the participants were also given a narrative account – to give them a mental image of the person they were guessing about. They were told that they were estimating the chances that a forty-five-year old man called Jack was an engineer rather than a lawyer. Jack, the participants were informed, is married and conservative, and enjoys carpentry and mathematical puzzles. Although this information is irrelevant to the statistical likelihood of Jack being an engineer or a lawyer, at least from a 'frequentist' probability perspective (i.e. probabilities calculated on the basis of how often an event occurs across a large number of trials), most people were excessively distracted by it. After being told Jack's story, they overestimated the chances that Jack is an engineer. De Neys and Goel wanted to capture how people were thinking about this Engineer-Lawyer task. They brought thirteen people into their lab and asked them to try the task while in the fMRI scanner. The experiment produced some fascinating results. Areas of the brain usually thought to be associated with System 2 analytical thinking (usually used when people are solving a statistical problem) did not dominate. The fMRI evidence picked up stronger activations in the emotional areas, suggesting that the participants were being distracted by the narrative information. They were using more subjective and emotional styles of thinking to resolve what was meant to be a mathematical problem.

Neuroscientific evidence is growing about the various ways in which our social instincts underlie a wide range of real-world decision-making.[16] Can we apply similar tools and insights to those used by De Neys and his colleagues to unravel self-interested herding and collective herding? Helping to answer these questions, neuroscientists and experimental psychologists are joining with economists to advance the new subdiscipline of neuroeconomics.[17] The types of neuroeconomic collaborations vary. Sometimes the economists provide the theory, models and analytical structure around which the neuroscientists build their own models. Sometimes the neuroscientists provide the economists with new tools to test innovative theoretical hypotheses, and this is where economics and neuroscience combine in the study of herding.

I first came across neuroeconomics at the American Economic Association annual meeting in Philadelphia in 2005. Before then, in my thinking about what happens when we are copying others I'd struggled, as many economists do, with the problem that the brain is a black box. After attending the session on neuroeconomics it occurred to me that perhaps neuroeconomics could fill a gap in economists' understanding of herding and anti-herding. After discussions with distinguished neuroscientist Wolfram Schultz and his team, based at the Department of Physiology, Development and Neuroscience at the University of Cambridge, we decided to combine neuroscientific techniques with economic insights to investigate herding.

Schultz was one of the pioneers in what was then the very new science of neuroeconomics. He is interested in how we learn, and particularly in how our reward pathways enable us to learn from the errors we make. His seminal contributions include the theory of *reward prediction error*.[18] This hypothesis links to reinforcement learning: the general idea that we and other animals learn to repeat actions when we associate those actions with reward. Animals learn because it is physiologically rewarding. Reward prediction error develops this idea but with an additional subtlety: animals learn behaviours not because of the direct stimulation they get from a reward, but because of the *errors* they make in their *prediction* of a reward. These prediction errors are picked up by neurons emitting the dopamine neurotransmitter (a chemical messenger) into reward-processing regions of the brain. For example, when a monkey randomly presses a lever and is surprised by the reward of a piece of fruit, the dopamine neurons emit a positive signal, encouraging the monkey to repeat the action. As she does so, she is again rewarded but is less surprised by the reward. Her reward prediction errors get smaller and smaller as she learns to predict more accurately the likelihood of a reward. When the prediction errors reach zero, the monkey's prediction of a reward and the actual rewards she receives have matched up, and learning stops.[19]

How can we connect this with our decisions to copy and herd with others? As we have seen, herding can be explained as the product of social learning, and social learning is driven by reward learning too. Together with Christopher Burke and Philippe Tobler (both now at the University of Zurich), Wolfram Schultz and I brought together economic and neuroscientific tools and insights in a neuroeconomic study of herding and social learning.[20] When people follow others their neural reward system is activated, but which neural areas specifically – those more usually associated with logical thinking or those more usually associated with instinctive emotional responses?

For our first experiment, we recruited a group of people comprising students and other adults from the local community around Cambridge. We asked them to decide whether or not to buy a financial share. If they made the right choice then they could earn some money. They were given some information to help them decide. In the first stage of the experiment, we gave the participants some private information in the form of a share price chart. In the second stage, we showed them the decisions of a herd – depicted in an image of four other people's faces – with a tick or a cross to denote whether the person had decided to buy or not.[21] To capture the social condition, we also showed our participants a photo of the faces of four chimps. Why? Generally, scientific experiments are controlled. To get an objective measure of how the experimental conditions are changing behaviour, a controlled experiment needs a baseline – and the control condition serves this purpose. For our fMRI experiments, our control condition needed to be similar to the human herd image, because otherwise any differences in the brain activations we measured when we introduced our experimental participants to the social information about the herd's choices might have been driven by differences in the visual stimuli, not by social influences (pictures of faces are more stimulating than no picture at all). The monkey faces were as close as we could get to human faces – but we did have to assume that our participants were unlikely to let a herd of monkeys dictate their financial choices. Then, using fMRI, we scanned the participants' brain activity as they were assessing the information and making their choices. We were curious to know what happens in people's brains when they are balancing private information and social information. When our participants were balancing private and social information, what neural mechanisms would be activated, not just for copycats herding but also for contrarians anti-herding?

We identified herding choices in two situations: first, when the participant decided to buy a share after seeing information that most of the herd (i.e. three or four out of four) had bought it too; and second, when the participant decided not to buy a share after seeing information that most of the herd

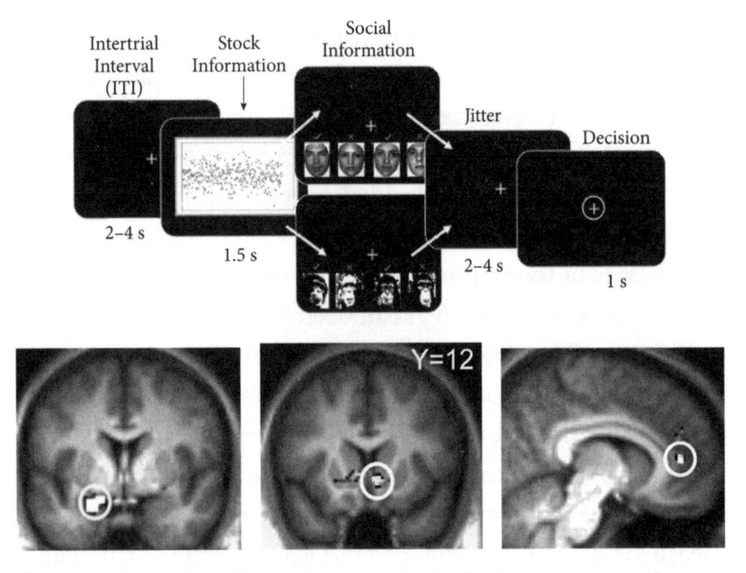

Figure 4. Financial herding and anti-herding in the brain scanner: task structure and brain activations in amygdala, prefrontal cortex and anterior cingulate cortex respectively.

had not bought it either. Contrarian anti-herding choices were identified in the opposite situations – when a participant bought the share even though most of the herd *had not* bought it, or when she did not buy the share even when she could see that most of the herd *had* bought it. We also analysed the impact of some of the participants' individual differences, which we captured by asking them to complete some biographical questionnaires and personality tests before we brought them into the brain scanner.

A strong herding tendency was identified in our participants. They were copying the herd's decisions far more often than we would expect if they were just deciding randomly. This confirms evidence from a diverse range of sources, that humans have a strong tendency towards herding; anti-herding is much more unusual. We are copycats much more often than we are contrarians. In order to pick up what was going on in the brains of our experimental participants we focused our analysis on those brain regions commonly implicated in decision-making. One of these is the amygdala, part of the limbic system (a collection of brain areas associated with emotional processing) and thought to be involved when we are processing negative emotions, including fear. Another is the ventral striatum, an area implicated in the processing of rewards, the focus of Schultz's reward prediction error model. Finally, we looked at activations in the anterior cingulate cortex, an area generally associated with higher cognitive functioning. There is some evidence that the anterior cingulate cortex operates something like Plato's charioteer: it steps in to resolve neural conflicts, including in situations where System 2 reason and System 1 emotion are competing.[22]

To explain what neuroscientists mean by a neural conflict, we can look to other studies from social neuroscience. A classic neuroeconomic study of social conflict was conducted by American neuroscientists Alan Sanfey, Jonathan Cohen and their colleagues.[23] The team brought nineteen people into their lab and asked them to play the ultimatum game, a famous experimental game widely used by behavioural economists to capture people's social preferences – that is, people's propensities to be selfish or generous.[24] As with many variants of this experiment, Sanfey and his colleagues split their participants into two groups. They gave one player (the 'proposer') $10 and asked them to divide the money between themselves and a second player (the 'responder'). If the responder accepts the proposer's offer, then the money will be allocated accordingly. If the responder rejects it, however, then neither player gets any money. The challenge for the proposer, then, is to figure out the lowest possible offer that the responder is likely to accept. Standard economics, at least if starkly presented, predicts that an offer of $1 should do it. If both players are rational, selfish maximisers, then the responder would not reject an offer of $1 when the alternative is $0, because a rational economic decision-maker will always prefer something to nothing. In contrast to the predictions of mainstream economics, however, many experiments with the ultimatum game show that proposers are surprisingly generous, and will offer not much less than 50 per cent of the total, while responders will reject relatively large offers, even if those offers are much greater than $1. This is interpreted by many as evidence of our socialised natures. Our propensity towards generosity means that proposers are inclined towards 'fairer' offers – where fair is defined as something approximating a 50:50 split. When responders decide that proposers are making unfair offers, they will punish the proposers by vetoing the offer even though the veto leaves the responder with nothing too.

Sanfey and his colleagues were interested to see how their responders would behave if they decided that they were being treated unfairly. They scanned the brains of the responders, targeting three main regions of interest: the insula, parts of the prefrontal cortex, and the anterior cingulate cortex. Neuroscientists think that the insula is implicated when we feel negative emotions like disgust. Disgust can be understood not only as physical repulsion, such as the feeling when we smell a foul odour. Disgust also has a social corollary: the disgust we feel when being unfairly treated. The experimenters found significant activations in all these brain areas. When their proposers offered the responders much less than half the money, activity in the insula captured the responders' social disgust of feeling cheated. This emotion was so strong that the responders were inclined to punish the unfair proposers by rejecting their mean offer, even when it meant getting nothing themselves. Sanfey and his colleagues also inferred that the prefrontal cortex was driving more economically sensible choices. On purely economic grounds, it is better to win a small sum than nothing at all. The anterior cingulate cortex was acting as arbiter, reconciling the conflict between the cognitive desire for more money and emotional responses such as anger and resentment that trigger retaliation when a person feels wronged.

For our experiment, we used the fMRI scanner to capture a different dimension to these social emotions – specifically the types of cognitive or emotional responses that drive us when we are herding or rebelling. With regard to the cognitive dimension, we hypothesised that our participants might be driven by a form of self-interested herding – linking to the Bayesian social learning experiments of Anderson and Holt that we introduced in chapter 1. Their findings were consistent with the idea that herding copycats are using Bayes' rule to reconcile contradictions between private information and social information. In our experiment, the private information was the objective evidence communicated via the share price charts and the social information was conveyed via the images of the choices of the herd.

We studied the fMRI evidence and identified significant differences in brain activity between the first phase, when the participants were looking at the share price chart (their private information), and the second phase, when they were looking at images depicting the herd and their choices (their social information). When the participants were looking at the social information about the herd's choices, areas of their ventral striatum were more strongly activated than when they were not looking at the social information. This finding is consistent with the idea that social information triggers reward learning. Non-social factors were important too, specifically the different participants' preferences for different types of stocks. There were two broad types of participants – some who preferred stocks with high average values, and others who preferred stocks with low average values. Why would someone prefer a stock with a low average value? Post-experiment questioning revealed that some participants thought that a stock with a low value today might turn into a high value stock in the future. Either way, both groups were predicting the likely rewards from the stocks and the ventral striatum was activated more strongly when participants were buying the type of stocks they generally preferred.

The activations in other brain areas differed depending on whether our participants were herding or anti-herding. When participants were herding, they showed significant activations in the amygdala – an area, as noted above, associated with processing negative emotions such as fear. This finding is consistent with the idea that herding and fear are somehow related. Perhaps when we are feeling fearful we want to avoid risks, and are thus more likely to collect together in groups and conform with the herd. We also found significant activations in the anterior cingulate cortex of the contrarians making anti-herding choices. One possibility, similar to Sanfey and his colleagues' interpretation of their brain scanning evidence, is that the anterior cingulate cortex is mediating a neural struggle. The ventral striatum is capturing our desire for reward, the amygdala is capturing the fear associated with the risk of disagreeing with the herd, and the anterior cingulate cortex is mediating this neural conflict.

As well as picking up some emotional processing, our experiments also captured how people respond to private and social information. Our evidence linked to two of herding's facets: first, how people use social versus private information; and second, how reason and emotion interact when they are balancing these different sources of information. Our fMRI evidence was consistent with the idea that a mixture of objective, cognitive and subjective, emotional influences was driving decisions to join the herd. Our finding links with insights from neuroscientists Ramsey Raafat, Nick Chater and Chris Frith, based at University College London. They identify the transmission of thoughts and information between individuals as a key characteristic of human herding, and suggest that interactions between unconscious 'automatic contagion' and conscious 'rational deliberation' drive this facet of herding.[25] Our experiments also illustrated something about the interplay between the rational, economic influences associated with economists' theories of self-interested herding and other social scientists' theories about the emotional drivers of collective herding. How can we link this experimental evidence with Kahneman's dual systems model? If Bayesian explanations are true, then whether information is private or social shouldn't matter. We process it all using higher cognitive functioning, drawing on our System 2 thinking. But then, why do we see activations in the emotional processing areas when people are thinking about what others in a herd are deciding? Paralleling the interpretation of the Engineer-Lawyer fMRI evidence about System 1 thinking from De Neys and colleagues, if Bayesian explanations are only part of the story (not necessarily false, just not the only thing going on), then we will see activations in areas usually associated with emotional, intuitive decision-making. The fact that neural areas associated with emotional processing are activated during herding suggests that it is not all about cool, calm calculation, despite what many economists might claim.

Herding heuristics

As we have seen, neuroeconomic experiments can capture interactions between emotion and cognition, connecting with Kahneman's division of System 1 thinking and System 2 thinking. Another insight from Kahneman's model links to the *speed* of decision-making. As we noted above, System 1 often dominates because it requires less cognitive effort. When we herd, is this because System 2 is lazy and we want to avoid the time and effort it takes to do the careful reasoning when there are quicker decision-making tactics available? If so, then following the herd will not be a controlled, logical choice. Instead, it may be a quick, automatic response driven by System 1. This connects with another set of insights from Daniel Kahneman about simple cognitive tools known as *heuristics* – quick decision-making rules. Perhaps in our herding experiments we were picking up the operation of a *herding heuristic*.[26]

How do herding heuristics work in practice? We herd because it is quicker and easier just to follow others, even if there is a chance we are simply copying their mistakes. Imagine you need to buy a new fridge, and you know that your neighbour has just spent a lot of time investigating the best brand of fridge to buy. Why would you repeat all that effort when you could just ask them for a recommendation? Your heuristic is to ask your neighbour. This will save you time and energy. But the problem with heuristics is that, whilst they are quick and convenient and often work well enough, they *sometimes*, though not always, lead to systematic mistakes – what behavioural economists and economic psychologists call *behavioural biases*. When we follow others, we may be leveraging valuable social information, or we may just be repeating their errors. Our neighbour may have bought their fridge on impulse, perhaps just because their neighbours had bought the same, without properly checking its specifications. If we follow them, then we too might end up with a second-rate fridge.

Cognitive psychologists Amos Tversky and Daniel Kahneman identified three main groups of heuristics and related biases: the *availability heuristic*, the *representativeness heuristic* and *anchoring/adjustment*. When we use the availability heuristic, we judge the chances of a specific event happening according to how easy it is for us to retrieve and recall relevant information. When we see a crowd in front of us, it is a clear and salient signal. Our vision of the herd in front of us is readily available and close to the top of our minds. By looking at the herd and simply copying them, we can circumvent other more costly cognitive devices which require more time and effort, for example memory or calculation. When we use the representativeness heuristic we are judging the likelihood of an outcome by comparing it with what we interpret as similar experiences and events in the past. This encourages us towards herding because we will assume that others' decisions provide clues as to how we should judge a situation ourselves. When we use anchoring and adjustment heuristics, herding will emerge if we are using the group's consensus as a social reference point. Again, this saves us time and effort because we don't have to start from scratch each time we are faced with a new choice. For example, if we are buying or renting a house, we may choose an area where members of our family or friends have recently moved. We use the information they give us about prices, local amenities and transport links as a reference point and, taking account of our own preferences too, we adjust our own choices around this reference point.

The psychologists Gerd Gigerenzer and Daniel Goldstein emphasise that using these heuristics is not irrational. Using heuristics is sensible. We build them into our routines to save us the time and energy that we would need to exert if we thought deeply about everything we do. For copycats, imitation is a 'fast and frugal' heuristic. It is a cognitive short-cut helping us to make quicker and more efficient decisions in social situations.[27] When we have social information we can be more selective. We can bypass laborious information-gathering exercises. Buying a fridge is a relatively simple choice – we want a machine that chills and freezes. For more complex purchases, we might need more help. For example, if we have a few friends and family who have recently bought new smartphones or computers, we may use a herding heuristic to guide our purchases. This might be more sensible than copying our neighbours' fridge choices, as noted above, because most of us can easily grasp the basic functions of a fridge. In the case of phones and computers, however, the information and options available are often overwhelmingly confusing. So, it makes sense to look to other people we know, who might know more than we do. We copy their purchases because we assume that they are knowledgeable, have researched what has the best capabilities and what is the best value, and can understand all the esoteric technical information about the various options. We couldn't do a better job ourselves in deciding which phone or computer to buy, so we rely on our herding heuristics to speed up and simplify what might otherwise be a complex, time-consuming and confusing decision-making process. Herding helps us to deal with the problem of choice and information overload – both of which might otherwise paralyse our decision-making.

All this raises the question of why and how we developed our herding tendencies in the first place. Where do our herding instincts come from, and do they suit the modern world? What connections are there with our basic instincts, developed over millions of years of evolutionary history? Evolutionary biology offers some key insights – not only about the evolutionary origins of dual-thinking systems in humans and other animals, but also the ways in which our instincts to herd might have served important purposes in primitive environments where resources were scarce. The quick, automatic and instinctive System 1 styles of thinking and deciding are older in evolutionary terms, whereas System 2 thinking, associated with conscious, deliberative, cognitive effort, has evolved more recently.

We have seen that there are number of ways in which we can apply Kahneman's division of System 1 fast and System 2 slow thinking styles in reconciling the different conceptions of self-interested herding and collective herding explored in chapters 1 and 2. Neuroeconomics adds to these insights by providing some evidence that our herding decisions involve complex interactions between cognition and emotion. They cannot be categorised, in any binary way, as either rational or irrational. Our choices to follow others may reflect a mixture of logical decision-making and more unconscious and emotional influences.

Fast interconnectedness in our modern globalised world affects our daily lives like never before. Technologies such as social media connect us closely together with complete strangers, sometimes many thousands of miles away. Information, goods and services, food and addictive substances are abundant and easy to come by. We can rapidly choose to consume something new with a click of a mouse button. This strange new world has enabled our basic copying instincts to spread on a massive scale. We can follow the social information collected via online reviews on Airbnb, Uber, eBay, TripAdvisor and price comparison sites, amongst many others. Whether we can or should trust these sites as much as we would a friend illustrates some of the limitations of copying behaviours when they emerge on an epidemic scale. When all our interactions are so anonymous, information can be manipulated to encourage us to follow fake news and other spurious information about what other people choose or think. Another worrying implication is that collective herding can dominate self-interested herding far more easily because modern technology suits our fast System 1 thinking style. When all our decisions can be made so quickly, there may be no time for the slow and careful reflection associated with System 2 thinking.

We have yet to explore the origins of these strong herding tendencies. We can look to behavioural ecology for evidence of the propensities towards self-interested and collective herding that we share with our animal cousins. And we can also look to evolutionary biology and evolutionary neuroscience to understand better how these responses have developed through our own evolutionary history. We shall explore these perspectives in the next chapter.

4

Animal herds

From the African plains to the Arctic tundra, huge numbers of animals all over the planet herd together to travel long distances. These animal herds are almost constantly on the move, escaping seasonal fluctuations in the weather and searching for new sources of food and water. Wildebeest, for instance, move together in enormous herds, often a million strong, as they make their 1,800-mile trip from the Tanzanian Serengeti to the Kenyan Masai Mara in the north, and then back again, chasing rainfall and fresh grass. Staying in one place risks death from starvation or thirst, but migration is also perilous. The wildebeest have to cross crocodile-infested waters and navigate other dangers. By herding together, the wildebeest balance the threats. The herd provides protection and increases the wildebeests' individual chances of survival, at the same time helping to ensure the survival of the species.[1]

We share with other animals a surprisingly wide range of similar instincts to herd in groups. Why have we and many other species developed such a strong and symbiotic relationship with others around us? In this chapter, we shall explore insights from behavioural ecology and evolutionary biology

Figure 5. Migrating wildebeest: herding together to survive the crocodiles.

to discover what lessons we can learn from the animal kingdom about the social instincts we share with many of our animal relatives.

Leopards versus wolves

Previous chapters have explored the differences between self-interested herding and collective herding. Just as our interactions with the groups around us are determined by differences in our characters and inclinations, so too are the social interactions between other animals. Sometimes animals are solitary. Sometimes they form coalitions in groups, for mutual benefit. Sometimes individual animals sacrifice themselves for their group and/or species.

Taking a closer look at the characteristics of leopards and wolves will illustrate some of the contrasts between the different types of human behaviour that we first outlined in chapter 1. There, we saw that economists often assume that our worlds are populated by a special type of individualistic, self-interested and self-contained agent – *Homo economicus. Homo economicus* can act independently of others because they are coordinated not by direct social interactions but by the Invisible Hand of the price mechanism. Given that herding is ubiquitous in the animal kingdom, especially amongst mammals, it is hard to think of any animal as self-contained as *Homo economicus*. Snow leopards might be the closest example. Fewer than 9,000 survive in the wild, and they are solitary creatures, living their lives in the mountainous alpine wilderness to ensure their own survival, without much contact with others of their species except when reproducing and defending their territories. Very few of us inhabit such solitary lives.

More sophisticated economic models capture the benefits we gain as self-interested individuals by cooperating and collaborating with others. We can learn from the herd, be protected by the group, or gain something tangible from collaborating with our fellows. Altruism plays no role in this. With self-interested herding, each individual animal prioritises its own needs and desires by collaborating with groups to ensure a better outcome for themselves. Such coalitions are common in the animal kingdom. When wolves hunt as a pack, for example, each individual wolf benefits from the coalition they join. Like the hunters in Rousseau's stag hunt game, introduced in chapter 1, wolves can catch much bigger prey when they operate together.

Linking with some of the insights about herding heuristics that we explored in the previous chapter, some behavioural ecologists have shown that pack dynamics are characterised by simple heuristics. Cristina Muro and her research team created a computer simulation in which wolf avatars hunted virtual prey. The only information available to the virtual wolves was the location of the prey and the other wolves. The simulation incorporated two basic rules: the wolf avatars would not risk their lives, only moving towards the prey as close as was safe; and once each wolf had reached this safe distance, it was programmed to move away from the other wolves. Apart from this, all the wolf avatars were identical and autonomous. Muro and her team discovered that the patterns generated by the computer simulations closely mirrored the behaviour of real wolf packs. Previous studies had suggested that complex hierarchies and forms of communication enabled wolves to hunt effectively. Muro's research suggested something simpler. In coalitions, wolves use simple rules that further their own interests, and in the process act in the interests of the wolf pack. If all the wolves hunt effectively together, then each individual wolf will benefit.[2] The lessons can be extended to self-interested herding, supporting the idea that human coalitions may also be using simple heuristics and rules of thumb to ensure they gain as individuals from the collaborations they develop with others, with concomitant benefits for the group as a whole.

The social lives of penguins and dragons

Self-interested herding is seen in other animals too, not just mammals and sophisticated pack hunters. In fact, herding through social learning is endemic in the animal kingdom.[3] Just as economists have suggested that we watch others when deciding which restaurant to pick, behavioural ecologist Étienne Danchin and his team postulated that animals glean important social clues about where to find food and mates from watching and copying what other animals are doing.[4] We described one specific example of this social learning in the introduction: quolls avoid eating poisonous cane toads because they have been taught this behaviour by their parents, or observed it in other quolls. But social learning is not confined solely to mammalian and marsupial species.

The Adélie penguins of the Antarctic are stuck in the middle of the food chain: they eat krill and in turn they are eaten by leopard seals. A penguin hunting for food risks being hunted and eaten itself. Social learning is the individual penguin's best strategy: each penguin waits to see whether the other penguins jump into the sea or not. Eventually, one penguin who is brave or hungry enough to take the risk makes the first leap. The other penguins watch to see how this penguin fares beneath the waves before judging if they should jump into the ocean too. If the first penguin survives, then the others herd behind, their collective behaviour determined by the prior penguin's fate.

Animals we might usually think of as less sophisticated, such as lizards, also share a knack for social learning. An international team of researchers led by Anna Kis from the University of Lincoln studied bearded dragons living in the deserts of Australia. By observing fellow lizards, the bearded dragons learnt to retrieve food by opening a trapdoor – a relatively complex cognitive task.[5] Just as the penguins and lizards got a good meal after engaging in a process of social learning, so a similar process is at work when people choose restaurants: we infer something about the quality of different restaurants from observing other people's choices.

Angry birds

Self-interested herding also provides protection, as when crossing a busy road with a large group of other pedestrians rather than singly. This herding has two dimensions – animals copy each other, and they group together – and we can see examples of both used by animals to escape predators. The simplest types of copying for safety involve camouflage. The dusty dottyback, a copycat reef-fish living in the Indo-Pacific coral reefs, is able to change colour quickly to mimic surrounding fish, and this helps to reduce detection by predators.[6] This in itself is not herding – but a similar effect of visual camouflage is achieved when many animals come together en masse. Predators struggle to focus on one target when lots of targets are gathered and moving together. Behavioural ecologists explain this as a *dilution effect*. The individual prey sought by predators is diluted within a large herd, making it hard for the predator to pick off lone targets easily. Within a herd, individual animals are less vulnerable.

Animals form coalitions, not only because packs of animals are better hunters together than alone, as we have seen, but also to protect themselves. Groups of animals can defend themselves effectively when they consciously work together. Meerkats, for example, are often observed taking it in turns as sentinels, watching out for danger.[7] Black-headed gulls and other birds form coalitions to warn other birds about risks, and sometimes come together in a mob to attack predators.[8] Domestic cats have experienced these tactics, including our cat Hobson. When he immigrated to Australia, his first experience of an antipodean backyard was not much fun. Hobson was spotted by a lone myna bird, whose piercing squawks soon drew five more birds and they all swooped down on him in formation, Hitchcock-style, and scared him indoors. The lone myna bird had instigated an impressive and clever coordinated attack.

Herding cows

These more concrete and objective benefits underlying self-interested herding run alongside unconscious influences encouraging us to join others in groups. We have seen that there are many sensible reasons for humans and other animals to gather together in groups and herds. These motivations for herding clearly help each selfish individual animal to survive, and so can be explained as a considered choice, consistent with Daniel Kahneman's analysis of slow System 2 thinking styles.

Other forms of herding are not so easy to explain directly in terms of either System 2 thinking or survival chances for the individual. This brings us to collective herding, which, as we explored in the previous chapter, seems to be more consistent with Kahneman's System 1 thinking, driven by instinct, impulse and unconscious motivations. We succumb to peer pressure, and experience intangible psychological satisfactions from our sense of belonging with others, even when we can see no clear, objective purpose to joining in a group – at least not from the perspective of our own self-interest.

If someone asked you to think of an animal herd, there is a good chance that you would think of cows. Cows are the archetypal herding animals, but they do not herd together out of blind stupidity. Cows are highly social animals with complex social hierarchies. They exhibit signs of stress when separated from their herd and they form strong bonds with other individual cows – in much the same way as humans identify single individuals as their best friends. In one study, behavioural ecologists measured the stress levels of cows by recording their heart rates and blood levels of cortisol (the 'stress hormone') in two scenarios: when the cows were put in a pen with an unknown cow, and when they were penned with their 'best friend'. Cows showed much-reduced signs of stress when they were with their friends.[9] What the cows experienced holds more generally, too, across most mammalian species. Like humans, many mammals feel less stress and more psychological satisfaction when they collect together with others they know.

Evolutionary influences

One of the mysteries of herding is why some individuals herd and copy others when it is not obviously in their best interest. Evolutionary biology can help to explain this anomaly because it does not focus on the individual animal, or even groups of animals. The selfish individualist is just a bit-part player when wider evolutionary imperatives are at stake. Whether herding is conscious and self-interested or unconscious and collective, it has evolved to maximise the chances of survival, not of the individual but of the species as a whole.

Charles Darwin's 1859 magnum opus *On the Origin of Species* provides a starting point in understanding the evolution of social (and anti-social) instincts in the animal kingdom. Different species have evolved characteristics that give them an adaptive advantage, helping them to thrive in their natural habitats. If they survive long enough to reproduce then the whole species is more likely to survive too. If the environment changes, however, then some species will die out because they no longer have an adaptive advantage in the changed environment.

Evolutionary biologists develop Darwin's ideas about natural selection to explore the different ways in which our outward behaviours have evolved in response to environmental constraints and obstacles. If our behaviour evolved a very long time ago, then it is not surprising that we do not always consciously understand why we behave in the ways we do. To better understand what drives us, we can make a distinction between *proximate causes* and *distal causes*. Proximate causes are the incentives and motivations that determine our day-to-day choices. We enjoy some foods more than others because the foods we prefer tap more effectively into the physiological systems that process our perception of reward. Distal causes explain the ultimate cause of our behaviour in evolutionary terms, as manifestations of our species' evolutionary fight for survival. To explain the difference between these proximate and distal causes, we can turn to the example of sugar. Many of us eat too much of it. We buy and eat sugary foods because we find them satisfying. Sugar causes physiological changes that trigger rewarding bodily sensations within us, and if our bodies signal that something is rewarding then we are more likely to want more of it. This is the proximate cause of our tendency to overeat sugary foods. The distal causes are not about our immediate, day-to-day, visceral responses. They are much older, and link to ancient mechanisms which evolved in our species hundreds of thousands of years ago. We evolved to forage for sugary foods because this helped us to find sufficient nourishment in a primitive world where nutritious, energy-full food was hard to come by. Enjoying and effectively digesting ripe fruit motivated us to find the rich energy sources scarce in primitive environments. We also evolved to store this energy as fat because, in primitive environments, we might have had to wait a long time before we found new sources of nutritious food. Those who liked and got enjoyable sensations from sugar were more likely to eat sugary foods, lay down fat stores and, when the famines came, survive to reproduce. These ancient mechanisms are the distal cause of our love of sugar.

How does adaptive advantage manifest itself in human and animal behaviour? Both self-interested herding and collective herding can be explained in terms of evolved mechanisms that served, and perhaps still serve, important purposes in increasing our chances of survival.[10] Herding is a form of adaptive advantage and it has distal causes. These distal causes reinforce the proximate causes that have implicitly formed the foundation of our analysis of herding so far.

In modern contexts, we learn to associate herding with reward and we are consciously and unconsciously motivated to join with others because we find it satisfying in some way. Self-interested herding is rewarding because it helps us to get what we want. Collective herding gives us less tangible, more unconscious psychological satisfactions, but these are just as crucial. Most of us enjoy being with our friends and family. Most of us are happier being part of some sort of group. We are more likely to join with other people and enjoy their collective support and safety. Whether motivated by self-interest or more diffuse and less conscious rewards, these are the proximate causes of self-interested and collective herding.

The distal causes reflect the value groups had in helping our ancestors survive in difficult primitive environments. Herding, whether self-interested or collective, is an inherited, innate strategy that we still use today. By observing and copying others our ancestors developed the best strategies for foraging, escaping predators and finding mates.[11] Our ancestors adapted to their environment using herding as a strategy to increase survival chances. They went on to reproduce and so passed on these herding instincts to their descendants, and thus our herding instincts evolved.

This evolutionary perspective also suggests that adaptive advantage is what self-interested herding and collective herding have in common. From an evolutionary perspective, both forms of herding are as much about increasing the chances of survival for the group and species as they are about helping the individual. Reconciling collective and self-interested herding from this perspective of evolutionary advantage also allows us to see that, for humans, both forms of herding reflect our social instincts and inclinations. We evolved our sociability and the common (but obviously not universal) aversion to aggression because, in this way, our ancestors' small communities had stable social structures and were better able to survive. Conformity served an important purpose in the evolution of our social instincts and herding tendencies, but in today's social media-saturated landscape, this conformity is perverted by overconnectedness. Conformity has been magnified far beyond what used to make evolutionary sense in primitive environments.

Self-sacrificing slime moulds

In evolutionary biology, the self-sacrificing individual is dispensable to its species and does not get a chance to reproduce its genes. We might think that cooperation and self-sacrifice are phenomena seen only in sophisticated animal species, reaching their apotheosis in humans. In fact, both cooperation and self-sacrifice are observed in relatively primitive life forms too, for example in the slime mould species *Dictyostelium discoideum*, a form of social amoeba. Different slime mould cells will cooperate even when they have different genotypes (different combinations of genes). This is unusual because most multicellular organisms are composed of cells from the same genotype, which makes evolutionary sense. From the perspective of the survival of the fittest, cells of the same genotype do not need to compete for resources because whether the cells survive or their genetically identical clones do, either way the genotype survives. The priority is the survival of the genes, not the individual cells. Slime moulds are unusual because they cooperate even when they do not share genes. So the survival of cells with one genotype is at the expense of cells with another genotype.[12]

In slime moulds, what form does cooperation take? Evolutionary biologist Paul B. Rainey studies *Dictyostelium discoideum* and has developed some interesting ideas about how and why slime moulds cooperate. Slime mould cells live in soil where they feed off the bacteria released by decaying leaves and animal droppings. In good times, each individual cell takes the form of a single-celled amoeba and moves around randomly, hoovering up bacteria. Sometimes, however, the environment throws up challenges. Nutrients become scarce. Then, chemical signals in the amoeba trigger a process of metamorphosis. The individual cells aggregate and self-organise to become a multicellular slug. Some of the amoebae metamorphose into the slug's stalk cells. Other amoebae form spores at the tip of the stalk cells and these are quickly released into the environment, ready to thrive when environmental conditions improve again. The stalk cells are not so lucky – they wither and die. The mystery here is why a single-cell organism would sacrifice its own reproductive chances to form the stalk cell of a multicellular organism – a dead-end in evolutionary terms. The stalk cells' genotype may die out at the same time as the stalk cells because these cells have no chance to reproduce. What if the negative environmental changes turn out to be temporary? In this case, each slime mould cell would have had a better chance of survival and reproduction if it had remained as a lone amoeba. Paul Rainey postulates that the single-cell slime moulds are evolutionarily programmed to balance risks. If they do join the other cells then they may land up as stalk cells. But, if they are luckier, they may form part of the slug that can reproduce via the release of spores. Rainey argues that the fate of the self-sacrificing slime mould cells is essentially bad luck.[13] Some slime mould cells are winners, others are self-sacrificing losers. Perhaps we share more in common with slime moulds than we might imagine.

Slime moulds illustrate the point that cooperation sometimes emerges not because the individual animal benefits, but because it helps a species to survive. Ants, too, are a highly cooperative species and they exhibit similar behaviours to self-interested human herding, as we explored in chapter 1. Ants also engage in social learning. The economist Alan Kirman was interested in the connections between animals' social behaviour and economic theory. He drew on observations from entomologists who had noticed that ants do not forage evenly across different food sources. When they are choosing between two sources of food, armies of ants tend to focus intensively on one or other of the sources. To explain this phenomenon, Kirman developed an 'ant model' of social learning. Kirman argues that the ants' copying behaviour is a manifestation of their recruitment activity. A single ant discovers a new source, they transfer this knowledge to other ants via an exchange of chemical signals, and in this way ant armies are recruited to forage one food source to the exclusion of another. There are benefits for the ant group if one food source is exploited more intensively than another because, by cooperating, the ants can forage more effectively. Eventually, the armies of ants will switch to the other source, perhaps when the first source has been depleted sufficiently. This social coordination helps the whole ant colony to survive. Apparently anomalous ant behaviour, difficult to explain from one ant's perspective, has an explanation that links to survival for the entire ant colony.[14]

Sociable animals

High levels of sociality and social functioning are shared across the animal kingdom. The biologist E.O. Wilson described some exemplars of social behaviour, the *eusocial* animals, which are characterised by their social sophistication. Eusociality is seen across a diversity of species, including ants, bees, wasps, termites and naked mole rats.[15] The concept of eusociality links to the idea that groups are favoured over individuals. Eusocial animals possess a sophisticated social awareness and they share highly developed instincts for cooperation. Eusocial animals practise 'kin selection': individual animals sacrifice their own chances of survival in order to favour the reproductive success of their relatives. In the organisation of eusocial animals' communities more broadly, altruism is a powerful force. Eusocial animals form strong, sometimes monogamous pair-bonds. They share caring for their offspring not only with their partners but also with other adult animals. Eusocial animals live in extensive colonies populated by overlapping generations of individuals. Within these colonies, there is a division of labour across different tasks, some of which eliminate an individual colony member's potential for reproduction (the worker bee is a well-known example).[16] Each individual animal has no independent purpose, and the colony functions more like a single animal.

The concept of eusociality is fascinating from a social science perspective too. It takes us back to some of the descriptions of mobs and collective herding that we looked at in chapter 2 – including the influential work of the psychologist Gustave Le Bon. Le Bon used a biological analogy to describe mobs. He explained how mobs form as a human body forms. Like the cells within a living body, the individuals in the mob have no independent life of their own. For Le Bon, the mob is like a

> being formed of heterogeneous elements, which for a moment are combined, exactly as the cells which constitute a living body form by their reunion a new being which displays characteristics very different from those possessed by each of the cells singly.[17]

Le Bon's insights suggest a way to build a link between the psychological and the biological explanations for grouping and herding. The psychology that brings mobs together has its corollary in behaviours observed in eusocial animals, in which the individual animal has no identity of its own, in the same way that the individual cells of a body have no independent existence. The concept of eusociality can also illustrate the differences between collective herding and self-interested herding. Collective herding is not always and obviously in the individual animal's self-interest, but it does work well from the group's perspective.

Teaching orcas

As we saw above, gathering for safety is a self-interested choice, but animals herd together for other reasons besides self-interest. From the perspective of the whole herd, the safety of individual animals also increases the chances of survival for the group as a whole. In this way, collective herding overwhelms each animal's individuality. Large animal herds often have a nature that cannot be explained solely in terms of the individual animals that comprise them. Like Le Bon's human mobs from chapter 2, the whole herd is something different from the sum of its parts. Wildebeest are a case in point. Individually timid, a herd of a million wildebeest gathered together makes an impressively loud noise.[18] It is a frightening and powerful force, with a large and independent nature of its own.

Social mammals also give us two examples of sophisticated social behaviours: teaching and culture. Orcas are one example. Orcas live their lives with their families, in 'pods'. Yet, like humans, female orcas stop reproducing in mid-life. In evolutionary terms this is a puzzle. What evolutionary explanation might there be for older orca females to live such a long post-reproductive life? International teams of behavioural ecologists thought that post-reproductive orcas might be able to teach us something about human menopause, until recently thought to be simply an otherwise inexplicable modern artefact of advances in public health and medicine.

Orca pods form matriarchal hierarchies. One much-studied pod is the J Pod, living in the Salish Sea, a network of waterways off the west coast of southern Canada and the northern United States. The J Pod was headed by the female orca J2, aka 'Granny', who had been studied by teams of behavioural ecologists ever since she was first photographed by Dr Ken Balcomb in 1967. She was the oldest orca known to humans, and is believed to have died in 2016 at an impressive age, possibly a hundred years old. She was in excellent health until her last sighting, in fact appearing much fitter than many much younger males. She probably had her last calf in her thirties or forties; certainly, she was never observed with a calf of her own during the last four decades of her life.

Behavioural ecologists think that orcas like Granny who live long post-reproductive lives play a range of crucial roles in orca society. One common theory is the grandmothering hypothesis, an idea used to explain why human females live so long after they have lost their capacity to reproduce. Older females without infants of their own can help younger females rear their offspring, increasing the survival chances of the whole group. In human populations, for example, there is evidence that children with grandmothers are more likely to survive longer.[19]

There are social learning explanations too, consistent with some economists' models of self-interested herding and social learning, but with an additional twist. Older orca females retain important social and environmental knowledge, and they *teach* this to the younger orcas, helping them to learn how to navigate their hunting grounds. This sharing of information is not just about younger orcas watching older orcas. Behavioural ecologists define teaching in terms of an individual incurring some cost to themselves in the process of imparting knowledge to others.[20] Teaching is more sophisticated and complex than learning. Learning just requires one individual to observe another, and an animal being observed by a social learner is passive, not necessarily encouraging or even noticing that another animal is learning by watching what they do. Teaching, on the other hand, is a consciously cooperative process. Both teacher and student are actively engaged in the process of sharing information and knowledge. Behavioural ecologists also note that teaching involves a level of self-sacrifice. The teacher incurs 'opportunity costs'. While they are teaching, they lose the opportunity to spend their time and effort looking after themselves, instead sacrificing their own interests to help another animal. Teachers may not benefit at all as individuals. Teachers do, however, help the group and therefore the species, so, from an evolutionary perspective, teaching serves an important social purpose. Teaching is certainly what Granny seemed to be pursuing in her later life. Lines of orcas would follow her during their salmon hunts. When Granny noticed younger orcas deviating from the path that she had set, she would hit the water with her tail, warning them to follow her. It was an interactive process.

The role played by older orcas in teaching and social support is complex and nuanced. Researchers noticed that Granny's bond with her son was particularly strong. Male orcas have a much shorter lifespan than the females, living to just thirty or so years whilst females commonly live beyond eighty years old. Like Granny, the surviving older female orcas in the J Pod also spent much more time with their sons than with their adult daughters. They shared salmon with their sons but not with their daughters, perhaps reflecting some ecological form of cost-benefit analysis. Supporting a son's reproduction is less costly than supporting a daughter's reproduction because sons mate with orcas from other pods, and those other pods carry the cost of the sons' calves, so it makes sense to give sons preferential treatment. If a son survives for longer, then he is more likely to reproduce, and when his calves are born they will not be a drain on his mother's pod's resources. On the other hand, if a daughter survives to reproduce then the pod will bear the costs of raising her calf. The researchers' actuarial calculations show that orca sons with living mothers survive for much longer than those without. When a mother dies, the mortality risk for her surviving son increases eightfold, whereas for a daughter it is much less.

Animal cultures

Culture is another phenomenon driven by our herding instincts and observed in numbers of social species, not just humans. Some of our evolved strategies reflect the evolution of culture over long periods of time.[21]

The sociobiologist Richard Dawkins has worked on extending Darwinian evolutionary concepts into the social realm, as in his path-breaking 1976 book *The Selfish Gene*. In this book, Dawkins asserts that Darwinian principles operate beyond genes and in the social world too. Memes – the human social equivalent of genes – are the ideas that move between us all, via language for example. Memes are the essential building blocks of our social interactions. They are the ideas and norms replicating via a process of *memetic contagion* through cultures and societies.[22] So, copycats are essential to this process. Although Dawkins' views are controversial amongst modern scientists[23] – the extent to which our destinies are formed by social institutions as well as our genetic makeup is a matter of dispute in sociobiology and evolutionary psychology – there is a basic consensus on how

Darwinian principles of natural selection also apply in the social world.

Cultural traditions form to bind societies and communities together, and cultural conformity helps species to survive. Herding plays an essential role in the transmission of culture and, in turn, culture helps to mould the social norms that reinforce animals' instincts to herd and imitate. Cultural norms also form the social structures against which contrarians can rebel. Cultural norms have been observed in a number of species, including whales and dolphins.[24] Behavioural ecologists have found that chimpanzee populations acquire local traditions in foraging for ants. Some chimps will use small sticks to collect a few ants at a time, eating the insects from the sticks. Other chimps will use a long stick and wait for many more ants to accumulate and then scoop them all into their mouths with their hands.[25] Behavioural ecologists believe that different styles of ant-eating represent different forms of chimp culture.[26]

Andrew Whiten, a psychologist from the University of St Andrews, devised an experiment to test whether cultural norms would spread through groups of monkeys. He and his colleagues studied 109 vervet monkeys living in the South African province of KwaZulu-Natal. In the first stage of the experiment two separate groups of monkeys were fed corn dyed different colours. One group was fed pink corn spiked with bitter leaves, and unspoilt blue corn. The second group was fed the opposite: their blue corn was spiked with bitter leaves and their pink corn was naturally appetising. The first group learned to prefer blue corn, the second to prefer pink corn. To capture whether the monkeys had learned from others, the researchers then observed the behaviour of twenty-seven baby monkeys born to the original monkeys. This younger generation was not exposed to any nasty-tasting corn. The researchers had given them the opportunity to enjoy both pink and blue corn, neither spiked with bitter leaves. So for the baby monkeys, they had no reason to favour one colour of corn over another – all the corn, whether pink or blue, was equally palatable. Even so, the baby monkeys copied their mothers in favouring just one colour of corn, either pink or blue.

So far, all of this is consistent with the social learning models we have already explored. But then the researchers noticed something else as well. Ten male monkeys moved from one group to the other. Monkeys who had been brought up in the pink-corn-preferring group moved to the blue-corn-preferring group, and vice versa. These migrant monkeys very quickly acquired the cultural norms of their new group and shifted their preferences away from one colour to the other. These monkeys had not tasted spiked corn and had not been taught by their mothers to avoid a specific colour of corn. There was no obvious objective reason for these monkeys to change their preference from blue to pink corn, or vice versa, other than the social influence of the other monkeys around them. The researchers attributed the monkeys' switch towards conformity with their new community to the power of cultural norms. Parallel phenomena have also been observed in humpback whales – with whales copying feeding traditions used by other whales, even though these were no more effective as hunting strategies.[27]

Cultural differences have been observed in other – 'lower' – species too. To assess the influence of cultural differences in a more controlled way, behavioural ecologists have studied the migration routes and schooling patterns of a species of fish called the French grunt. The schooling behaviours observed in different populations persisted beyond the grunts' lifespan. To understand why, the researchers took individual fish from one population and moved them to another population at a new site. Using their social learning skills, the new fish quickly adopted the traditions of their fellows in terms of feeding sites and migration routes. More interestingly, this experiment also allowed the scientists to exclude the possibility that these foraging traditions were a product of environmental or genetic factors. When the fish were moved to a new site but were given no opportunity to observe the behaviour of the population of fish there, they did not adopt the same foraging patterns, but instead developed their own.[28] The researchers concluded that the copying behaviours were not simple instincts, formed in response to the characteristics of resources available at different sites. They were social traditions paralleling humans' different cultural norms and traditions, and driven by the same types of copying and herding behaviours.

The evolution of human herding

We have seen that evolutionary biology illuminates the social instincts that we share with other animals. So what are the key differences if both self-interested herding and collective herding have adaptive advantages in common? Evolutionary neuroscience provides us with a potential explanation, and can tell us more about humans' evolved social instincts, including our instincts to herd.

Modern humans, *Homo sapiens*, evolved around 200,000 years ago and were characterised not only by their opposable thumbs and upright posture but also by their large brains. According to some neuroscientists, our social instincts paralleled the evolution of our brains – which some biologists attribute to our high levels of sociality, a characteristic shared with other mammals.[29] Evolutionary neuroscientists postulate that our brains have three distinct parts, each representing different stages in our evolutionary development. The brain stem is a remnant of our reptilian brain, the limbic system is a remnant of our mammalian brain, and the neo-cortex (of which our prefrontal cortex is one component) is an evolved feature of modern hominid brains.[30] This schema is controversial. Some neuroscientists argue that evolutionary models of the brain are too simplistic. This simple idea is powerful, however, in suggesting that our behaviours reflect an interaction of primitive and sophisticated responses, each driven by different neural areas with different evolutionary histories.

How does this link to herding? Older, less evolved brain areas are common across the animal kingdom from lizards to apes, and associated with more instinctive, primitive emotional responses – including some of the System 1 fast-thinking styles. Perhaps these ancient impulses link to some of the unconscious motivations driving collective herding. Areas concentrated in our neo-cortex – associated with deeper, more logical thinking including high levels of cognitive functioning and sociality – have evolved more recently, alongside the evolution of our System 2 slow-thinking styles. These might explain tendencies towards self-interested herding. If so, then self-interested herding and collective herding may just be different forms of adaptive advantage, developed at different stages in our evolutionary history. Perhaps they are similar survival strategies, triggered by our evolved cooperative instincts and predisposing us towards joining and imitating groups.[31]

The importance of being docile

We have seen how animals, including 'lower' life forms such as slime moulds, sacrifice themselves, but why have self-sacrificial instincts evolved in humans? Herbert Simon, whose role in developing theories of heuristics we noted in the previous chapter, was also keen to explain why some people are more self-sacrificing than others. He thought that we could use the phenomenon of self-sacrifice to develop a better understanding of the evolution of pro-social instincts. Simon postulated that social groups work better when they include altruistic individuals who are conformist and suggestible in the face of group pressure. He formulated a mathematical model to show that these altruists are beneficial in evolutionary terms. Without a minimum proportion of altruists within our populations, our species cannot survive.

How did Simon explain his claim? He started by delineating a specific personality trait that self-sacrificing conformists share – what he called *docility*. Docile individuals are super-receptive to social influences. They have an emotional intelligence that enables them to learn from social information quickly. Docility is a form of social heuristic – a quick decision-making rule, linking to the herding heuristics we explored in the previous chapter. Docile individuals will believe many things without needing direct proof, and this enables them to absorb social information quickly and easily. We might argue that the docile people in our populations will be easily exploited by mendacious non-docile individuals, fostering tyranny and oppression. Simon was more optimistic, holding that docile people might be essential in helping groups to survive environmental challenges. The presence of docile conformists gives human populations an adaptive advantage, boosting our fitness for survival.[32]

The interesting thing about Simon's concept of docility is that it is not in an individual's selfish interests to be docile. Docile people may be good at assessing social information, but they are not doing it for their own sake. Herbert Simon's concept of docility suggests that there are some psychological characteristics playing specific roles in ensuring the survival of social species. Simon's model of docility illustrates that humans, alongside other mammals, have evolved as highly social species, and some of this can be explained via insights from evolutionary neuroscience.

Theory of mind

Our highly evolved social instincts, including our propensities to copy and herd, are linked to our capacity for *theory of mind* – that is, the inferences we make about the beliefs, feelings and actions of others. When our brains process social information, our responses may be formed not just from our direct experience of watching others, but also by our empathetic and imagined emotional responses. Neuroscientists have discovered that when we imagine other people's experiences, particularly those close to us, our neural responses are much the same as if we'd experienced the events ourselves. Empathy has evolved to help us, as social animals, to understand and share emotions.

Neuroscientific studies show that these empathetic responses engage automatic emotion-processing circuits, some of which evolved long ago in evolutionary time. Tania Singer and her colleagues at the Wellcome Department of Imaging Neuroscience at University College London conducted one such study. Singer and her team invited sixteen twentysomething couples to their lab to participate in an empathy experiment. With their neural responses being monitored using fMRI brain scanning, all the participants were given mild electric shocks, and the women were also asked to observe the shocks being inflicted on their partners. The experimenters found that neural networks for pain were activated not only when the women themselves were being shocked but also when they saw their partners experiencing pain.[33] One explanation for this could be that we have evolved to respond emotionally to the suffering of our close family and friends. Partly, this serves a learning function. In the process of empathising with others' discomfort we can predict the consequences for ourselves of a similar experience.[34] Our neural responses mirror the responses we would have if we were experiencing the same pain that we are observing in others. The researchers inferred that this empathy engages automatic, emotional processing mechanisms in areas such as the insula – a relatively old area of the brain, associated with the processing of a wide range of 'valenced' emotions, that is, emotions that have both a positive 'good' dimension and a negative 'bad' dimension. Negatively valenced emotions include fear, disgust and sadness. Positively valenced emotions include trust, love and happiness.

How do these behaviours link to the evolution of our own instincts to herd or rebel? As we saw in the previous chapter, neuroscientific evidence suggests that herding choices might reflect an interaction of System 1 quick thinking and System 2 slow thinking. System 1 and System 2 may also interact when our instincts to imitate are driven by a theory of mind.[35] Social emotions such as empathy play a role. We imagine ourselves in someone else's position and this allows us to understand what they are thinking and feeling. For example, a good teacher will put themselves in the mind of their students and imagine what confusion might be like for them. Perhaps teachers draw on their own earlier experiences as a student themselves, and then pitch their lesson to suit. We also use theory of mind to help ourselves. This process of 'mentalising' can help us to deal with situations in which information is unclear and incomplete. When driving on the motorway, for instance, our instincts for self-preservation motivate us to do all we can to avoid an accident. We use our high levels of social functioning, including our theory of mind capacities, to put ourselves in the mind of other drivers and drive accordingly, anticipating the decisions other drivers will make and so avoiding a crash.

How do our social instincts, such as theory of mind, link with our neural functioning? Neuroscientists have identified a specific area of the brain known as Brodmann area 10. This brain area is implicated in our ability to mentalise about the beliefs and actions of others. Significant activations of Brodmann area 10 have been observed in people playing games involving trust, cooperation and punishment. Deficits in this area are thought to have links with autism, a neuro-developmental disorder associated with theory of mind constraints. Children on the autism spectrum, including those with Asperger syndrome (a milder form of autism), do not easily understand emotions and social cues.[36] These limits on social comprehension seem to be associated with relatively high activations in the ventral prefrontal cortex for people with mild autism. Perhaps this suggests that people on the autism spectrum realise that their ability to empathise is constrained, and so exert cognitive effort in attempting to overcome this deficit.

Monkey mirrors

Theory of mind can explain why humans and monkeys have evolved common instincts to copy others. What drives these high levels of social functioning? More and more neuroscientists are studying how brain structures drive the high levels of social functioning that have evolved in social animals. What is actually happening physiologically when we copy others? Some neuroscientific studies have identified motor responses associated with empathetic inferences about others' pain, for example using transcranial magnetic stimulation studies, which involve activating specific brain areas via temporary magnetic stimulation.[37] Specific neurons – von Economo neurons (sometimes known as spindle neurons) – have been identified by other economists in humans as well as monkeys, apes, whales, dolphins and elephants and are implicated in humans' and other higher mammals' social capabilities.[38] Single neuron experiments on primates have captured a similar type of neuron linking our imitative instincts – the mirror neurons. In single neuron experiments, an electrode is inserted into a single neuron and measures the electrophysiological impulses passing through it. If these impulses are strong, then it can be inferred that the neuron is being used intensively. Mirror neurons are found in the pre-motor areas of the primate brain – less evolved areas than the prefrontal cortex, and not under primates' conscious control. When a monkey observes another monkey engaged in an action – for example, grabbing a banana – then the observer monkey's mirror neurons are activated in much the same way as if they were grabbing the banana themselves.[39]

Systems of mirror neurons – that is, *mirror systems* – have evolved in humans as well as monkeys, but their function is still a subject of speculation. By mirroring others' behaviour perhaps we can implicitly understand copycats' emotions and actions. This helps us to predict what drives others, and we can use this information to improve our own decisions. But it is difficult to get direct evidence of mirror system activity in humans because single-neuron experiments are extremely invasive. If we can infer from the primate experiments that human imitation reflects the same mirroring processes as detected in monkeys, then we have a potential link with herding. In humans, imitation learning mediated by mirror systems may be connected with the sophisticated social forms of learning, associated with phenomena such as language and culture.[40] Mirror systems may also explain herding through social learning, one of our key explanations for self-interested herding.

Vulcans in a social media world

As we have seen through this chapter, many of our copycat behaviours are the outcome of evolutionary forces from millennia before we invented the modern technologies dominating our lives today.[41] These evolved behaviours helped us to survive in the small social groups characteristic of primitive hunter-gatherer settings. They also helped us to learn more effectively, because in small groups, individuals were better able to observe and monitor their peers' behaviour. Emotions, for example impulsivity, helped us to survive in harsh natural environments where basic resources were often scarce and perishable. Quick action was essential to avoid starvation. So, limbic structures in the brain evolved to encourage impulsive emotional responses, including impulsive collective herding.

Imitation allows good ideas and important information to move quickly through species of copycats.[42] A similar phenomenon is observed at an emotional level. In both monkeys and humans, emotions can travel fast. Emotional contagion is observed in children when they cry, and in adults when caught in disaster scenarios. Mourners' emotions – for example, those felt by the throngs that gathered outside Buckingham Palace after Princess Diana's death, as described in the introduction – are another example of how emotions spread in mobs and crowds. These epidemic emotions may serve as an important survival mechanism. Emotions are driven by System 1 thinking and we process them quickly. Emotional contagion is beyond our conscious control and spreads through crowds involuntarily. Waves of emotion can rapidly wash over a group giving each individual a signal, for example to flee or fight. In this way, emotional contagion can help animals to survive by allowing rapid, unconscious responses without requiring any conscious coordination by any single individual – useful in emergency situations. Neuroscientist Ramsey Raafat and his colleagues have suggested that emotional contagion specifically, and social contagion more generally, may have evolutionary value because they enable emotions to ripple quickly through crowds of copycats, reinforcing societal norms.[43]

Are our primitive evolved instincts to copy and follow a problem? In simple hunter-gatherer communities, the likelihood of divergence between individual and group interests was small. Any individual exhibiting deviant behaviours would be quickly noticed and ostracised or excluded. Over the course of human history and with the growth of civilisation, however, individual and social interests have diverged, a divergence which has intensified with the rise of the twentieth- and twenty-first-century technologies, especially those associated with computerisation and globalisation. These have profound implications for our daily lives, but they have developed in a millisecond relative to evolutionary time. We have not had chance effectively to adapt our behaviour, including our herding tendencies, to modern institutions like markets and government, and modern artefacts such as money and computers.

Neuroscientist Jonathan Cohen takes an optimistic perspective on this tension between our evolved instincts and our modern world. Social influences are an essential aspect of our brains' evolution. When we lived in smaller groups, the chances of repeated interaction were greater. As our social instincts evolved, we developed strong emotional responses to selfish and exploitative behaviour, and these protected us. Cohen's view is that instinctive responses evolved for a purpose, and that even if that purpose has now been lost, the apparent misfit between our evolved behaviours and our technology-driven world may not be as destructive as we may fear. This is because our brains are 'vulcanised' – just as rubber can be vulcanised with sulphur to harden it and make it more resilient. Vulcanised human brains, according to Cohen, have evolved into a confederation of mechanisms, mostly cooperating but sometimes competing. The evolution of our prefrontal cortex has given us some resilience, allowing us to moderate the power of emotions across a range of decision-making domains.[44]

The distal causes of self-interested herding and collective herding may be similar, but the proximate causes are driven by different neural mechanisms, each developing at different points in our evolutionary history. Insights from evolutionary biology can help us to see that herding has evolved not to serve the purposes of lone individuals. From analyses of social animals we can see that self-sacrifice is a common mechanism, used by many animals to promote survival of the species.

How do these influences play out in our modern world? Destructive choices by one individual, or one small group of individuals, have seismically different impacts today than they would have had when *Homo sapiens* first evolved. In a primitive context, when tribes battled with each other, loss of life was small. Today, instincts for self-sacrifice favouring the interests of one group over another – such as in the context of global terrorism – can potentially have enormous and disastrous consequences for our species as a whole. At the extreme, countries with access to long-range weapons, including nuclear missiles, have the capacity to inflict death and destruction on a massive scale in the process of favouring their in-groups. Economically, globalisation, while allowing some groups to amass fortunes, has been associated with wide-ranging poverty and inequality for vast numbers of others.

Our evolved herding instincts can generate perverse outcomes in a technology-saturated world. New innovations have helped us to build virtual social connections around the globe, without the old-fashioned costs and sanctions that previously would have encouraged caution – though recent exposés of the unethical exploitation of personal data by some of these sites may change this landscape again. Social media have allowed the rapid transmission of information from copycat to copycat, so, in theory at least, we can now be much better informed about what is happening in the world from moment to moment. But have we become overconnected? Perhaps our extensive connections with different people around the globe magnify the dark sides of our copycat and contrarian natures? Fake news and cyber-bullying, funnelled through and facilitated by social media, mean that our evolved copying instincts spread very rapidly. The consequences are potentially enormous given the myriad interconnections between us enabled by modern technologies. What are the implications for herding and anti-herding today? To illuminate these tensions and how they play out in the modern world, the following chapters analyse the diversity of copycat and contrarian characters and the conflicts between them that we can see every day.

5

Mavericks

So far, we have seen the many ways in which our lives are dominated by convention and imitation. Mavericks and contrarians are in a small minority most of the time. They often have traits that are rare and unusual and that sometimes can seem strange, even sinister. Yet many of us are drawn to mavericks, perhaps because we think they encapsulate something lacking in our own personalities and inclinations. Or perhaps we realise that herds of copycats can't lead themselves, so we look to contrarians to lead us. We need contrarians to be in the minority, however, because too many of them would create chaos.

The psychosociologist and psychoanalyst Wilhelm Reich, whose ideas about crowd psychology we introduced in chapter 2, captures the spirit of the maverick, as well as some of our conflicted attitudes towards them. Politically, Reich was a radical. He joined the Austrian Communist Party in 1928. He advocated large-scale social change, including sexual liberation, as well as radical improvement in social conditions for the poor. His ideas were controversial and struck many as being strange and perverse. His personal life was chequered and complex. He was known – was infamous, in some circles – for his promotion of free love, allegedly coining the term 'sexual revolution'. He is most notorious for advocating orgasm as the solution to social and psychological problems, going so far as to invent the 'orgone energy accumulator', a sort of cupboard to facilitate orgasmic experiences. The accumulators came in a range of designs and finishes, from carpet-lined to egg-shaped. (Movie buffs may remember Woody Allen's parody of the orgone accumulator as the 'orgasmatron' in his 1973 science fiction comedy movie *Sleeper*.)

Like many mavericks, Reich had a roller-coaster career. A distinguished psychiatrist and psychoanalyst, he enjoyed early success at the heart of the Viennese psychoanalytical community – including a stint serving as deputy director of Freud's Ambulatorium clinic. Reich's early writings on mass psychology and character were also influential, inspiring new generations of psychoanalysts, including Sigmund Freud's daughter Anna. He even managed to persuade major figures to submit to his oddball orgone therapy as well as to test the theory. After extensive dialogues with Reich, even Albert Einstein was persuaded to conduct some experiments on orgone energy. But Reich's advocacy of controversial therapeutic techniques, including primal therapy and 'vegetotherapy' (the therapeutic use of massage as a form of release), attracted criticism from the press and his peers. From the 1940s on, Reich was forced to self-publish a lot of his idiosyncratic theories. Whilst he had many devotees, his ideas lost popular support. Caught up in the McCarthy-era surveillance and ostracism of communists, his maverick life ended pitifully, at the age of sixty, in prison, having been convicted of violating an injunction prohibiting the distribution of his orgone energy accumulators.[1]

Celebrated or derided, Reich's maverick ideas were for a long time at least tolerated. The tipping point for many mavericks like Reich seems to be when their ideas fail to complement what we already know, want, think or believe – when they are too much at odds with the prevailing zeitgeist. When a maverick's ideas lose all connection with reality as the herd understands it, the balance of public acceptability turns against them. The problem, also illustrated by Reich, is that this tipping point is not necessarily determined by a majority-view consensus. Vested interests who control public perceptions have always aimed to silence mavericks who rebel or who the group decides are dangerous. In today's world, social media give these vested interests much more power to circulate emotive messages widely and quickly.

Even so, the majority are often deeply suspicious and intolerant of mavericks. For example, societal conventions around marriage, the family and domestic life are often rigid traditions. Today, many women still risk their reputations, familial and social ostracism, physical isolation, psychological damage, violence and even threats on their lives when they aspire to make choices that sit uneasily with tradition. The journalist Upasana Chauhan, who wrote about her parents' violent threats when she was making her own choices about marriage, provides an example. Born in Haryana, India, Chauhan met and fell in love with a man from another caste. When she told her parents she wanted to marry him, their first response was to threaten to kill him. They put their daughter under house arrest until she convinced them that she would not marry without their blessing.[2] Upasana Chauhan was much luckier than many others in a similar position because, eventually, her parents and community agreed to support her marriage and husband. Other mavericks in similar situations have signed up for lives as social outcasts, and sometimes much worse.[3]

These struggles are one illustration of the conflicted feelings that mavericks inspire in us. We may worry about their intentions or methods. We may be confused by the ideas they are trying to convey. But at the same time, they fascinate us.

Mavericks' singularity is undeniably interesting. Indeed, many of us could call ourselves novophiles. In principle at least, we like what's new, different and unique. New ideas have value, and it is mavericks who are often the ones brave enough to abandon old ways in favour of new. As John Maynard Keynes observed, 'The difficulty lies, not in the new ideas, but in escaping from the old ones, which ramify, for those brought up as most of us have been, into every corner of our minds.'[4]

The illustrator William Heath Robinson tapped into our love of the unusual with his drawings of fantastical contraptions – some not too dissimilar to Reich's orgone cupboard. His inventions were whimsical and wacky, often held together with not much more than string. One of his ideas was the 'Wart Chair', a complex device designed to remove warts from heads; another a 'multi-movement' machine for gathering Easter eggs. He also imagined a device for killing flies with a ukulele.[5] Heath Robinson's name has gone into the English lexicon as an adjective to describe things patched together in a higgledy-piggledy, make-do way.[6] But his concepts were not just for amusement: his early illustrations were designed to counterbalance German propaganda and lift the spirits of troops suffering misery in the trenches of the First World War. His later drawings poked gentle fun at the pomposity and bureaucracy of experts, his machines a metaphor for Byzantine bureaucratic systems and structures in the interwar world. Maverick, sometimes even mad ideas can play a social and political purpose as long as they are well judged, in the collective subjective opinion. Sometimes we need more of them to counterbalance the excessive dominance of herding and convention in so many aspects of our lives. We like to see contrarians taking on authority and the Establishment. We admire contrarians' independence of thought. Their ideas are engaging and inspire our imaginations and optimism.

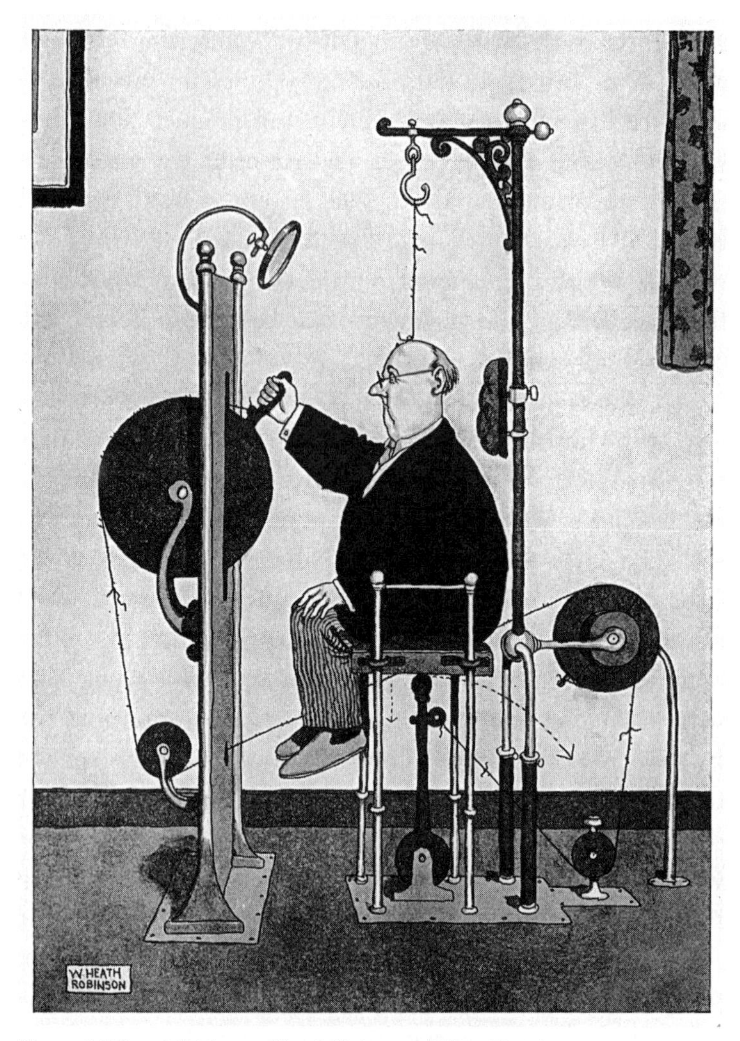

Figure 6. Maverick ideas: a Heath Robinson 'Wart Chair'.

Why be a maverick?

So – what motivates mavericks? What drives them to make their own way, risking social approbation in the process? And why do some of us choose a maverick path while others avoid it? Just like the social influences driving our conformist natures, contrarian behaviour is driven by a complex range of economic, social and psychological influences. In previous chapters we explored some of the behaviours that characterise copycats. Sometimes copycats are driven by self-interest, at other times by some sort of collective consciousness. Either way, individual copycats choose to move with a herd for a wide range of reasons, many of which link to their own welfare and survival chances in an uncertain world. It may seem that contrarians should be much harder to understand and explain; certainly, the literature on contrarians and mavericks is much smaller than that on copycats and herding. But in fact, mavericks' choices to rebel against a crowd can reflect a surprisingly similar set of motivations as those infuencing copycats. We are copycats partly because there are economic incentives to join the crowd, and these incentives tap into our self-interest. Mavericks are also propelled by self-interest. They use social information, they build their reputations, and they balance trade-offs between risk and reward. These are the corollaries of self-interested herding. Similarly, mavericks have incentives to promote their own individual advantage, but by acting *contrary* to the crowd. They are balancing the economic incentives too, but deciding on the opposite course of action. Their preferences incline them towards rebellion and dissent. Running alongside these consciously individualistic motivations are the corollaries of the unconscious drivers of collective herding. Often contrarianism is not the product of a rational calculation of relative benefits. Sometimes contrarians are motivated by psychological influences including cognitive biases, personality and emotions, as we shall see.

Mavericks and information

A good place to start our exploration of the incentives driving mavericks is with the economists' models of self-interested herding. These economic models suggest that, fundamentally, copycats and contrarians are not that different. Both types are rationally maximising their own self-interest – they just balance the incentives to come up with a different sort of decision.

The starkest economic models capture some aspects of mavericks and contrarians quite well. If information is good and uncertainty is limited then it makes economic sense just to get on and do your own thing. Sacrificing self-interest helps no-one when our choices and decisions are coordinated via anonymous markets and other institutions. But we do not live in this sort of world. We live in a world in which information is poor, uncertainty is endemic and market failures are everywhere.

As we saw in chapter 1 with the example of choosing between two restaurants, economic models of herding focus on the balance of private and social information. Economists postulate that we use mathematical rules, specifically Bayes' rule, to balance these different types of information. When choosing between two restaurants, we may have some private information – for example a friend's recommendation; and we may have some social information – one restaurant is crowded and the other is empty. The larger the crowd in one restaurant, the more likely that we will choose to eat there too. Why might a contrarian choose to eat at the other establishment?

In this simple restaurant scenario, private and social information are treated equally according to the quantity of the evidence. The different pieces of information are like signals. We have a private signal (the restaurant review or a friend's recommendation) and a number of social signals equivalent to the number of people already eating (we infer that each person has chosen that restaurant for a reason). For most people, the large number of social signals will outweigh the one private signals. Contrarians are more likely to over-weight their private signal. Some contrarians are very confident in their own power to decide well without worrying about what others are doing. They discount the social information implicit in the choices of others around them, weighting their own private signals much more heavily than ordinary mortals susceptible to persuasion by others.

By embedding these insights, economic herding models can be adapted to capture mavericks too. In 1998, herding model innovator David Hirshleifer and his PhD student Robert Noah adapted the herding model to 'misfits' – essentially capturing the behaviour of mavericks and contrarians. They argued that self-interested herding is disrupted by the presence of misfits – which is a good thing if the herd is going in the wrong direction. Misfits can play an essential role in social progress and improving social welfare, depending on the type of misfit.[7] In Hirshleifer and Noah's view, there is a range of types of individual who are inclined to eschew the queue. There are the Newcomers, who have had no chance to observe the herd because either they have only recently arrived, are not well placed to use social information and/or are prohibited from joining the herd for some reason. Then there are the Prophets, who have better private information (and know it) and so are less likely to be swayed by the actions of others. Joining the Newcomers and Prophets are the overconfident Fools, who do not really know better than others around them but believe that they do. Arrogantly, they falsely over-weight their private judgement and let it trump the social information conveyed in the actions of the herd. Then there are the Rebels, who have different payoffs – perhaps they get some additional satisfaction from rebelling itself, and so are more inclined to discount the social information implicit in others' choices.[8]

The problem is that all these types of contrarians – Newcomers, Fools, Prophets and Rebels – are behaving the same way by anti-herding. Just from watching them, we can't tell the difference between them because we have no information to judge how reliable they may or may not be. They all move against the herd, but for very different reasons. Some of them are contrarians for reasons that might not suit us, or could mislead us. How do we know whether or not we should follow them? The problem is that there is no clear solution. The uncertain herd might want to weight more strongly the information implicit in Prophets' choices and discount the actions of Newcomers and Fools. We might try to find out more about the contrarians to establish whether they have a reputation for reliability or prescience. Prophets who have a long track record will have built up a good reputation if they are truly wiser. Whether or not it is wise to copy a Rebel is less clear – we might decide we would like to copy them because we want to emulate their independent natures. We are conflicted: we like the thought of being unconventional but we do not want to be alone. We may lack the confidence to be a lone contrarian, but can be encouraged to join a small band of contrarians if a Rebel is prepared to take the lead.

Maverick risks

One characteristic that most mavericks obviously share is that they relish taking the risks that copycats prefer to avoid. This is easiest to see in the context of financial contrarianism, where maverick risk is an established practice.[9] Hollywood has popularised many examples, both fictional and real – from Gordon Gekko of *Wall Street*, an asocial criminal who has no regard for others in his strategies for making money, to the real-life maverick traders depicted in the 2015 biographic *The Big Short*, who displayed at least some social conscience as they hunted profits. This small group of mavens bet against the mortgage-backed assets created during the boom in the American sub-prime mortgage market during the 1990s and 2000s. Ridiculed and dismissed before the crash, they were proved right and made plenty of money out of their foresight, founded on their clever analysis of the objective evidence showing how unstable US subprime mortgage markets had become. More generally, however, mavericks have a lot to lose when they dissent. A speculator going against the market, for example by buying a financial asset when everyone else is selling it, is taking a big risk. They may lose a lot in terms of money, but they also risk their reputation if they are wrong. Why might a maverick speculator decide to risk anti-herding? Because, potentially, the rewards are very large for the speculator who can outwit the market. The risks faced by copycats and contrarians link to another economic model – a model of conformity developed by the American economist Douglas Bernheim. He explored the idea that conformity has value for self-interested individuals preoccupied with status, but for others contrarianism has more value. Bernheim argues that mavericks differ from copycats not only because they enjoy being contrarians but also because of their extreme preferences, manifested in the risks they are prepared to take in violating social norms.[10]

To understand how the risk dimension operates, we first need to learn more about how economists capture risk-taking. Economists have done a lot of work on risk. The standard view in economic theory is that risky choices can be captured by some embellishments of *utility theory*, one of the building blocks of mainstream economic theory. 'Utility' is the economists' word for happiness and satisfaction. We get utility from something if we think it is useful (where 'useful' is defined very broadly). According to the simplest versions of mainstream economic theory, we aim to maximise our utility from all the things we purchase and enjoy because we want to do the best for ourselves.

Expected utility theory brings the element of chance into the picture.[11] We do not know for sure what will happen next, and so we think about the relative chances of different things occurring: we form *expectations* of future outcomes. When we buy a lottery ticket, we balance our expectation of winning a prize against our expectation of not winning a prize. If we are forming expectations rationally, then we will know that the chances of us hitting the jackpot are very small and the chances of us getting nothing at all are very large (after all, why else would revenue-chasing organisations sell lottery tickets?).

To capture people's different attitudes towards risk, economists connect expectations with how our utility changes when we get more of something we like. As the saying goes, you can have too much of a good thing. The more we have of something, the less we enjoy some more of it. If we have eaten one chocolate bar, for example, we might quite enjoy a second, but the third chocolate bar – not so much. If we have eaten ten chocolate bars, then we are not likely to get much satisfaction at all from an eleventh. This illustrates the economic principle of *diminishing marginal utility*: our utility is diminishing with each extra chocolate bar we consume. Extreme outcomes – having either no or lots of chocolate bars to eat – don't bring us much extra utility. We prefer an average outcome – perhaps five or so chocolate bars.

Economists assume that, for most people, money is characterised by this diminishing marginal utility property too, and this links to risk. In economic theory, when a risk-averse person is offered a choice between a guaranteed sum of money – say $10 – and a gamble which gives them a 10 per cent chance of winning $100 but a 90 per cent chance of $0, they will avoid the gamble. This is because risk-averse people do not like extreme outcomes – the prospect of winning $100 is not appealing to them if at the same time they risk being left with nothing. Risk-averse people prefer average outcomes. They will forgo the chances of winning a large prize in order not to lose a lot. Risk-loving people have the opposite attitude – they have an increasing marginal utility for money. The more money they have the more they want. So they are happier gambling on extremes. They may lose everything but when they win, the utility they gain from these extra winnings will be magnified.

We can apply these ideas about expected utility and risk to our analysis of risk-seeking mavericks. Conformists prefer to be average because conforming means not much is lost, even if nothing much is gained. Contrarians, on the other hand, not only get less satisfaction from being conventional, they also get more out of chasing extremes because they enjoy the risk of being different. They want to move away from all the advantages that following the herd offers in terms of an averagely satisfactory existence. They want to take a chance on something different, even if they risk losing everything in the process.

Beating the crowd

Maverick risk-taking also connects with mavericks' desires to beat the crowd. Sometimes the winner takes all, and second-comers are left with little or nothing. The prize can take the form of money, applause or reputation. To beat everyone else, mavericks are often prepared to take extreme risks by investing a great deal, either in personal or in monetary terms. Invention and innovation illustrate well the vital importance of being first. Scientific researchers get very little credit for replicating other scientists' findings, even if the originality of their insight is failing to replicate bad results from another scientist's flawed research. Nevertheless, they are unlikely to get much attention. They will struggle to publish their contrary evidence because scientific journals are biased towards original and positive findings, and are not so interested in research suggesting that another researcher's original insights are wrong.

Beating the crowd links to reputation. Reputation is not unimportant to mavericks, but their reputation-building strategies are distinctive. As we saw in chapter 1, copycats' reputations are less vulnerable because they have ensured that when they are wrong, lots of other copycats are wrong too and in the same way. Mavericks take a different perspective on reputation. They prioritise their contrarian reputations so that, when they are right when others are wrong, they can reap large rewards. They build their reputations around being different rather than similar. Just as reputation can be protected by copying others' actions, so it can be enhanced when a person develops a new, original idea. Inventor of Post-it notes Alan Amron demonstrates the importance of reputation to trailblazers. In his battles with 3M over the provenance of his investment, Amron was concerned as much about 3M's claims that they had invented Post-its as he was about his $400 million financial settlement: 'I just want them to admit that I am the inventor and that they will stop saying that they are the inventor . . . Every single day that they keep claiming they invented it damages my reputation and defames me.' Amron lives in a social world, and so being recognised as the product's inventor was essential to his reputation and pride.[12]

Expected utility theory can explain only some of these links between anti-herding and risk-taking. Beyond standard economics, behavioural economists and economic psychologists have developed critiques of expected utility theory – perhaps most famously the psychologists Daniel Kahneman and Amos Tversky, as a background to their alternative theory of risk, *prospect theory*.[13] Kahneman and Tversky conducted some experiments that identified shifting and unstable attitudes towards risk, contrary to expected utility theory.[14] Expected utility theorists assume that people's risk preferences are stable – if someone is risk-averse then they are risk-averse, and simply reframing the choice will not change their minds. Against this, Kahneman and Tversky provided evidence that our preferences for risk-taking are determined by the way in which choices are framed, particularly in their concept of *loss aversion*. For an expected utility theorist, if someone is asked to take a bet on winning $10 versus winning nothing then they will make the same choice if they're asked to take a bet on losing nothing versus losing $10, because in either case, the difference between winning and losing is $10. Kahneman and Tversky's experiments suggested something different. Our risk preferences shift depending on whether we are deciding about gains or losses. We care much more about losing $10 than we care about winning $10. As Kahneman and Tversky succinctly describe it, for most of us 'losses loom larger than gains'.[15] How does this help us to understand the differences between copycats and contrarians? Copycats may be more concerned about what they might lose if they rebel than about the prospect of risk-taking. As we have seen in previous chapters, with self-interested herding each individual can collect information, find safety and power, safeguard their reputation and avoid the costs of not conforming: social exclusion and ostracism. Copycats worry about all that they might lose from rebellion. Contrarians may be less concerned about the losses they incur. They might actively invite being set apart from the crowd, or may be happier taking risks by deviating from social norms and hierarchies. This phenomenon is more consistent with Kahneman and Tversky's psychological analyses of risk than with economists' expected utility theory.[16]

Other biases may be driven by social comparisons, triggering contrarian responses. A US field experiment conducted by a group of economists from Harvard and Yale tested the impact of social comparisons on employees' contributions to their retirement savings. They gave employees information about the retirement-savings decisions of their peers. For the low-income employees, the outcome was unexpected:

providing information about some of their peers' ample retirement savings was associated with lower retirement savings for the low-income group. The researchers explained this behaviour as a contrarian 'oppositional reaction'. The relatively low-paid group did not want to engage with information that highlighted social comparisons with their richer colleagues. Information about their richer peers' choices just reminded low-income employees of their relatively low status, and so they resisted copying their colleagues.[17]

Maverick minds

Our different attitudes towards risk also connect to our personalities, as they do to demographic characteristics including age, gender and educational attainment. Experimental and anecdotal evidence confirms that contrarians do have the traits we would expect them to have: lower levels of risk aversion, lower levels of conformity and greater optimism – as measured via standard personality tests. Mavericks in the business world illustrate some of these traits. Often, business leaders and CEOs are expected to lead rather than follow. Evidence suggests that CEOs are also more likely to be risk-takers and are also likely to be good team-builders and optimists.[18] At a more personal level, business leaders often have distinctive qualities and attributes – many linking to maverick and antisocial tendencies – which may offend some and charm others. Entrepreneurial mavericks are not universally popular. Take the extraordinarily successful entrepreneur Steve Jobs. People who worked with him had very differing experiences of his personality: some thought him inspirational, others found him difficult and uncompromising. As one of his biographers, Karen Blumenthal, has observed, he was a man who thought differently.[19] He was a contrarian.

What is driving this maverick behaviour? What maverick psychology underlies the maverick personality? In chapter 3, we explored how different thinking styles can explain the different facets of copycats' characters. When we decide to copy others, sometimes it is more conscious and deliberate – consistent with Kahneman's System 2 slow thinking. Other times it is more intuitive, unconscious and/or instinctive – consistent with Kahneman's System 1 fast thinking.[20] These insights can be reversed for contrarians and mavericks. Some mavericks might thrive on the System 1-thinking physiological rewards associated with the buzz of taking risks, following your gut, doing something new and totally different.[21] These System 1 influences will operate alongside System 2 thinking: the deliberation and deep intelligence that the maverick taps into when developing new ideas and innovations. Mavericks may be consciously, deliberately taking risks and/or focusing on the future because they believe that their risks today will deliver rewards in the long term. For successful mavericks, there is a balance between the two. Instinctive risk-seeking is moderated by careful reflection in developing ideas and strategies.

If mavericks are driven by a System 1–System 2 interplay between emotion and cognition, then the standard economic theories of expected utility may need a rethink. The neuroscientific concept of reward better captures what mavericks are about. There is an extensive neuroscience literature on risk and reward, and much evidence has shown that dopamine pathways are involved in the processing of reward from risk-taking. These include the rewards we get from satisfying hunger, thirst, desire and other basic drives. Modern behaviours also engage the dopaminergic pathways – including overeating and drug-taking. Reward engages a complex series of neural structures implicated both in basic, instinctive emotional responses and in higher-level cognitive decision-making. For contrarian decision-making, similar interactions will play a role. Whether we are herding or anti-herding we are balancing the rewards against how we feel when we take risks. Some will enjoy risk-taking, others not so much. Perhaps the difference between copycats and contrarians is mainly that the latter viscerally enjoy risk-taking more than the former.

There are some nuances, however. As we saw in chapter 3, research into the neuroeconomics of herding reveals interesting neuroscientific data on activations in reward-processing areas of the brain when mavericks are anti-herding. Contrarian choices were associated with relatively stronger activations in the anterior cingulate cortex. As we noted in chapter 3, this is an area associated with higher levels of cognitive functioning, and so may suggest that contrarians are making cognitive effort to dampen down their ingrained impulses to follow the crowd. This would be consistent with the hypothesis that mavericks taking risks via their contrarian choices are not being impulsive in the way that risk-taking might be impulsive in other decision-making domains, such as gambling.[22]

Why we need mavericks

The maverick's incentives and motivations to rebel are clear from the perspective of the individual, but there are also some important implications for society at large. Mavericks can bring external benefits to the world around them. Mavericks may be independently minded, but this does not preclude them from pro-social desire to make a difference, be useful or inspire others. Mavericks can bring to the world new ideas and fresh approaches. Sometimes independence of thought is in harmony with the needs and choices of the herd.

Some mavericks can change our lives partly because they shift the balance of opinion. Cass Sunstein has explored some of the trade-offs between conformity and dissent. Most of us choose conformity as the most rational strategy, but from a wider perspective, conformity can lead society into big mistakes. Conformity sometimes reflects a lack of information but the problem of well-intended conformity driven by social learning is compounded because people are not always honest about what they believe and what they know. Most people's need to conform intensifies this dishonesty. In this way, Sunstein explains that widespread conformity exacerbates information gaps and encourages opportunistic behaviours associated with concealing information. Society's institutions can provide a partial solution. When democracies are working well, institutions such as the press and the legislature will help to ensure that we identify the truth. Institutions do not always work well, however, and then contrarian dissenters have an essential role to play, especially in the echo chambers of social media. Mavericks can be more honest and transparent because they care less about how people will respond to their dissent, and societies need dissenters prepared to resist social pressures.[23] Mavericks also serve important social purposes in challenging convention and preserving private information to ensure that it is not swamped by herds of conformist copycats chasing social approval. Mavericks and contrarians help to ensure that important information, ideas and principles are not lost to society at large.

Maverick dissenters have improved our social and political lives in many ways. But they can also create confusion, chaos and, at worst, destruction. Whom we label as maverick dissenters may itself be politically motivated. Different types of maverick change our world in different ways, for better and for worse. We can learn more about their impacts by looking at some specific types, from inventors through to whistleblowers.

Inventors

'Mad' inventors are the archetypal mavericks. They think laterally and are not wedded to how things have been done in the past. These instincts and abilities enable them to develop genuinely useful inventions. They are driven by their own intrinsic motivations to solve intellectual, mechanical or business challenges that they have set themselves. They do not always act in opposition to the crowd. Rather, they seem to act independently of the crowd.

Modern civilisation is characterised by the varied partnerships between inventors and entrepreneurs.[24] We have maverick inventors, engineers, chemists, physicists, computer scientists, biologists and medical scientists to thank for many of our everyday conveniences – everything from electricity, railways, antibiotics, computers and the internet through to can-openers and zips. In his fascinating account of some modern inventions, historian Gavin Weightman explains how the 'eureka' moments that gave rise to many of the inventions we take for granted today were in gestation for many years and sometimes decades. Sometimes this was because the ideas came from maverick amateurs who had the spark of originality but lacked the practical skills and knowledge needed to bring a product to market.[25] But none of the things that improve our lives significantly would exist if mavericks of one form or another had not come along and decided that we needed something novel and different.

Rebels

Rebels are the superstars of the maverick world, and their rebellious acts have been glorified – and vilified – for millennia. Over human history, our philosophical, religious and political lives have been driven by rebels, from Socrates and Galileo to Martin Luther King and Nelson Mandela. Rebels obviously share with other mavericks an independent nature, but unlike inventors and entrepreneurs they are driven by a desire not to produce new ideas, but to struggle against old ones. Given that the rebel's *raison d'être* is to be in contraposition to the rest, to go against the herd and to oppose convention, the power of their actions depends on the existence of a status quo to oppose. The status quo, which for copycats is a reference point, becomes an inverted reference point for rebels. They use the status quo to identify what they don't want, or don't want to be.[26] So, whilst rebels are motivated by a desire to act contrary to the crowd, they are not completely independent of it. Successful rebel leaders need a keen social intelligence and an awareness of the sentiments of the crowds around them. In this sense, they are as dependent on crowds as copycats, but in a different way. Without a crowd to watch, support and follow them, rebels have neither purpose nor much chance of success.

We need rebels because they have the capacity to change our world, sometimes for the better. By taking a maverick and contrarian view, these thinkers, activists and revolutionaries have not just been propelled by a desire to do something different and unusual. As important, possibly more important, is their willingness and capacity to do something to transform people's lives. We may judge some rebels to be good, evil or misguided, depending on our particular perspective. But, at heart, many rebels probably thought they were on a right and just path. And they are intrinsically valuable because they force the herd into that important balance identified by Cass Sunstein – between conformity and dissent.

History's most famous rebels have well understood their symbiotic relationship with the copycats following them. The Argentine Marxist revolutionary and cult hero Che Guevara was perhaps *the* archetypal twentieth-century rebel – and he exhibited all of the maverick traits we have described earlier in this chapter. His colleague Fernando Barral described him as being 'incredibly sure of himself and totally independent in his opinions. He was very dynamic, restless and unconventional . . . the most striking thing about him was his absolute fearlessness.'[27] But Che also recognised that his comrades were just as important as he was in supporting his rebellion against capitalist governments in Latin America. Che's own accounts suggest that he had a high degree of social intelligence, captured in his descriptions of the emotional impact he had on his comrades as a guerrilla doctor:

> in the early nomadic phase of guerrilla warfare, the guerrilla doctor must go everywhere with his comrades . . . He must undertake the exhausting and sometimes heartbreaking task of looking after sick men without having in his possession the medicine that would enable him to save a man's life. During this stage, the doctor has the most influence on the other men and their morale, because, to a man in pain, a simple aspirin takes on importance, if it is administered by someone who identifies with his suffering. During this phase, the doctor must identify completely with the ideals of the revolution, for his words will have more impact on the men than anybody else's.[28]

Che's social intelligence helped him to understand what motivates and drives people, how to secure his comrades' loyalty, and how to build solidarity with the revolutionary cause. Yet he never became part of the crowd. Even at his most selfless, when caring for and medically administering to his men, uppermost in Che's rebel mind was his individual impact and influence in leading, not following, the herd.

Rebels do not have to be famous revolutionary combatants to play important roles in social and political change. Sometimes, seemingly small acts of rebellion can have a large political impact. Though fashion is often dismissed as an ephemeral or trivial matter, historically, fashion statements have played crucial roles in political and social change – most strikingly in the context of women's rights. Amelia Bloomer, born in the United States in 1818, rebelled against the fashion constraints that (literally) bound women of the day. A leader of suffrage campaigns and influential in the women's rights movement, she demanded clothing for women very different from the tight corsets that then dominated women's fashion. When the women's rights activist Elizabeth Miller introduced loose-fitting trousers for women, designed to enable more freedom of movement and healthier living, Bloomer promoted them enthusiastically – and gave her name to them. Bloomers became not only a more comfortable alternative to the women's dresses of the time, but also a symbol of the women's rights movement. As is often true of maverick ideas and inventions, bloomers fell out of fashion – though the essential idea that propelled them to fame, that women should be enabled to live their lives more easily and comfortably, did endure, alongside the significant political and social changes associated with the emancipation of women.[29]

Whistleblowers

Whistleblowers are another type of maverick with a capacity for changing our world for the better. But, unlike rebels, they are reluctant mavericks. They do not create, but they do throw light on problems that the rest of us might be tempted to bury. They share the autonomous nature of other mavericks. Whereas rebels have an inherent, irrepressible instinct for rebellion, whistleblowers are more likely to be hostages to fortune. Sufficiently independently minded and principled, they are willing to call out the transgressions of others, but often do so hesitantly. In other circumstances, many whistleblowers might be happy just to blend with the crowd. They often act anonymously and off-the-record because they rationally fear the consequences of overt rebellion, even though, through their self-sacrificing actions, they can have significant impact on improving the welfare of others.

Whistleblowers have been crucial in challenging corporate fraud, political transgressions large and small, improper and dangerous medical practices, and physical and sexual abuse. They are indispensable in catalysing essential change and reform of financial, legal and healthcare systems. Yet societal and institutional attitudes to whistleblowers are often conflicted. Partly this reflects a form of short-termism. The rewards from whistleblowing take a long time to become clear. The media, politicians and society as a whole may have pressing short-term imperatives which mean that they do not want the fuss of the scandals catalysed by whistleblowers.

The ways in which we respond to mavericks demonstrate the negative social welfare implications of penalising a contrarian view. Whistleblowers are the most vulnerable of mavericks, and they often suffer severe penalties for their actions because the crowd is not necessarily inclined to welcome their dissent. Whistleblowers are often castigated for expressing a contrary opinion. Societal approbation reflects this short-term perspective. In the heat of the moment, we may jump to the conclusion that whistleblowers are nothing more than disloyal curmudgeons. In 2003, the United Nations weapons inspector David Kelly revealed to British newspaper journalists off the record and anonymously that he did not believe that sites he had inspected in Iraq were laboratories set up to manufacture biological weapons, as was the official line of the British and American governments at the time.[30] Allegedly against his will, he was publicly cited as the source of information undermining assertions about the threat from weapons of mass destruction.[31] He died soon afterwards. The official verdict was one of suicide, although doubts remain about the nature of his death.[32] Kelly paid a heavy price for his whistleblowing.

Unfortunately, David Kelly's experience is not so unusual. In recent years, numbers of high-profile whistleblower stories have hit the headlines. Some countries are starting to recognise the long-term consequences of vilifying whistleblowers, and new legislation and institutions are emerging to protect whistleblowers' rights and interests.[33] We need incentives to encourage whistleblowing, and protections for those who are made vulnerable by it – such as the setting of regulatory limits and systems to ensure that they are not penalised. The media has lionised a small number of whistleblowers, recently and most famously Julian Assange and Edward Snowden, each of whom have now been the subject of documentaries and Hollywood movies. But, caught up in the buzz of fame, do whistleblowers really help the crucial cause of exposing wrong-doing? In this, Snowden's elusiveness is more reassuring than Assange's celebrity.

Yet, despite these measures to protect whistleblowers, legislation remains difficult to implement if the consequences for those who speak out are irreversible. Added to this, the transgressions that whistleblowers are calling out are often dispersed across a number of different people, from perpetrators through to their allies, many of whom will have the motive and opportunity to conceal or destroy incriminating information. When evidence is missing, relevant authorities in the courts and elsewhere will not be able to identify precisely who is responsible and so the cases made by whistleblowers will be hard to prove.

We have seen that herding has many negative implications, especially in today's overconnected world. Mavericks bring benefits of their own in terms of new ideas and inventions, but they also restrain some of our copycat tendencies. In this context, mavericks are important to us because they counterbalance herding's negative consequences. In order for mavericks to rebel against the herd we might need additional incentives for people to take those risks; the problem then is in deciding whether or not rebels are on the right path. How can we agree on policies to encourage the 'good' rebels and discourage the 'bad' ones, especially when there may be little agreement about what is good and what is bad in our complex modern societies? Democratic institutions, including a free and unbiased media, can help us as citizens to make up our own minds about rebels and other mavericks with the power to change our lives. Much of the world, however, enjoys neither a free press nor other democratic institutions.

Mavericks also play a special role in the economy, reflecting the influence of two particular maverick types: entrepreneurs and inventors. We have already learnt something about the latter in this chapter, but in marketplaces, alongside entrepreneurs, inventors face some very specific constraints. Entrepreneurs and inventors may create their own reward in the sense of the enjoyment they get from building a new business or inventing a new gadget, but these activities are not cheap. Financing new business ventures and innovations is a struggle, especially for new and small businesses. This brings us to another side of the economy: alongside the entrepreneurs struggling to fund themselves, modern financial markets churn through trillions of dollars each day. How can entrepreneurs connect with some of this money? They must get past the gatekeepers of these global financial riches – creating a whole new set of problems for our copycats and contrarians. How can we ensure that speculators effectively channel money and finance towards the entrepreneurs and innovators with the best ideas for generating employment and economic success? We turn to these questions in the next chapter.

6

Entrepreneurs versus speculators

The economist and statesman John Maynard Keynes was a colourful, fascinating character.[1] His deep understanding of how economies work was formed by the very best education Britain could offer: he attended Eton College and went on to excel in his undergraduate studies at King's College, Cambridge. His intuitive understanding of speculators and entrepreneurs reflected more than his intellectual gifts and deep knowledge, however. Keynes was well placed to understand the workings of financial traders' minds because he traded himself – very successfully, all told, with a few memorable failures too. A possibly apocryphal tale from his Cambridge days is that he contracted to buy grain on forward markets. The contract date for the forward trade arrived before he had had a chance to sell his grain, and he was forced to store it in the King's College chapel.[2] Nonetheless, Keynes enjoyed considerable success with his idiosyncratic financial trading. Managing the college endowment as the bursar of King's College, he achieved excess returns over market averages of around 8 per cent. A key to his success was his focus on equities and stock-picking.[3] Keynes also had a good understanding of entrepreneurship. In chapter 12 of his 1936 magnum opus *The General Theory of Employment, Interest and Money*, he presented a powerful account of the psychology driving entrepreneurs to invest in their businesses. He was particularly interested in how uncertainty slowed them down.

Keynes was arguably the greatest economist of the twentieth century. Part of his genius reflected his impressive practical and intuitive understanding of real-world business. He understood how entrepreneurs and speculators are motivated by social as well as economic drivers. He also understood how the symbiotic relationship between them plays out in the macroeconomy. Entrepreneurs need finance to invest in building their businesses. Fast-moving, liquid financial markets work well in providing businesses with finance quickly. Keynes' enduring insights help to explain the 2007/08 global financial crisis and other episodes of financial instability. In a world filled with uncertainty, and when our conformist instincts dominate our financial choices, financial crises are not at all rare.[4] And the impacts are wide-ranging: without government intervention, the interplay between business and finance will not deliver what an economy needs in terms of employment and production.

Keynes was the first economist to explore the many ways in which social interactions between copycats and contrarians help and hinder business investment and finance. Financial markets have changed a lot since Keynes' time. The complexity of financial market interactions has grown with the advent of modern technologies, including algorithmic trading. The decision of one trader can precipitate large and volatile fluctuations, as lots of other traders can almost instantaneously decide to follow along behind. Nonetheless, Keynes' fascinating analyses embed enduring insights about the social incentives driving speculators and entrepreneurs to be the copycats and contrarians of the business world.

So, Keynes' analyses are a good starting point for illustrating how the business world is as prone to interplays between copycats and contrarians as any other aspect of our lives. Are successful entrepreneurs more likely to be mavericks? Why are speculators more often copycats? How do copycats and contrarians interact in the economy? In this chapter, we will answer these questions by exploring how and why copycats and contrarians respond to social influences in their pursuit of profit and new business opportunities.

The money convention

Money is our starting point in analysing the social interactions between speculators and entrepreneurs. Money unifies speculators and entrepreneurs, but in perhaps surprising ways. Generally, when it comes to money, most of us have a copycat side to our natures. We follow a money convention.[5]

How does the money convention work? Tangible forms of money – notes and coins – would be of no use to us if the rest of the herd were not prepared to take them as what economists call a 'unit of exchange' – in other words, we can exchange money for stuff, and our employers pay us money in exchange for our labour. Money has other purposes too, including its role as a unit of account. Accountants measure individuals' profits, losses, incomes and tax bills using money as their measurement unit. At a macroeconomic scale, statistical agencies use money to measure national income and output. All this only works because we have evolved the social convention of using money for our economic and financial transactions. We swap around our otherwise worthless bits of paper and cheap metal without thinking too hard about how and why this works. A Martian visiting Earth may well be puzzled by the value we place on certain types of paper and cheap metal. She may be even more puzzled by the fact that all we need to do is wave a bit of plastic at a metal box and we can take away carloads of groceries and household goods. Most of us are paid our wages and salaries electronically and see nothing directly tangible in return for our labour. We just follow the social convention that is money because everyone else adopts it too, and because our central banks and governments endorse and support it.

In our modern world, the money convention has become very complex, so complex that perhaps we have lost sight of money's essential purpose in terms of enabling economic activity by boosting production and employment. The globalisation of computerisation has enabled the emergence of innovative financial technologies in the form of new electronic money and crypto-currencies – in recent years, most famously Bitcoin. Bitcoin does not share all the features of conventional money, but there are ways in which it could replace conventional money. People have bought Bitcoin as a speculative opportunity and it could, in theory, be used as a unit of exchange and account, though so far at least, most of us are unlikely to have used it in our economic transactions. Until a Bitcoin convention is more widely adopted, it and other crypto-currencies will struggle to be anything other than a speculative curiosity.[6] Other alternatives to conventional money can be used in small communities, and some cities and suburbs have experimented with new, localised forms of money, such as the Bristol pound and the Brixton pound.[7] Essentially, these community-based money conventions complement the money conventions dictated by governments and central banks. If the Brixton pound were not somehow convertible into pounds sterling, directly or indirectly, then very few people would use it. So, overall, money is still a convention that relies on copycats to survive. Whatever sort of money we use, it only works if enough of the herd believes in it as a unit of exchange.

Our money convention is not silly. Even though money is an increasingly intangible instrument, it is nonetheless a clever and useful thing. Even old forms of money are economically efficient innovations. Before money we used barter, which is clumsy and involves very high transaction costs and search costs. In other words, it is inconvenient and time-consuming to use, especially when bartering something specialised and complex. Imagine, for example, you want to buy a new computer. In a bartering world, you would need to go and find someone with a computer they wanted to sell, and barter with them for something of yours they wanted to buy. Before the internet, you would have been confined to people you knew or living nearby – transport and travel costs would have been prohibitive. And even if you could find someone locally who wanted to get rid of their computer, you would have to have exactly what they wanted in exchange – an unlikely scenario. The chances of finding a neighbour willing or able to sell you exactly what you want are most of the time likely to be small. The chances that your preference for their possessions matches theirs for yours are even smaller. But in a world with money you can go to a shop, give the shop owners money that they can use to buy other things, and take away one of their computers in exchange. This explains why economies are more successful when they are populated by lots of copycats following a money convention.

Tulipmania

If money is a social convention then it needs a good proportion of copycats supporting it to succeed as a unit of exchange and account. Over time, however, money has morphed into something more. It has transformed into a way to make money out of money. Markets have evolved around the trading of money and other assets – and it is in these markets that the copycat speculators live. Financial assets tend to be relatively homogenous, what economists call 'fungible' – each unit is identical, and so can be swapped for another very easily. This makes financial markets very quick and liquid: they move fast and, superficially at least, smoothly. Speculators have entered this financial ecosystem to make profits from the trading opportunities available. Speculators move quickly, some would say impulsively.

Financial history shows us that herds of speculators are a powerful force driving financial markets and financial instability.[8] Indeed, speculative episodes are an enduring feature of financial markets.[9] Destabilising speculative fads and frenzies have been common throughout history. In recent times, not much more than a decade passes before a new one emerges, from the South Sea Bubble of the eighteenth century to the 1929 Wall Street Crash, the 1997 Asian financial crisis, the dot-com bubble of the 1990s/2000s, the 2007/08 subprime mortgage crisis and a series of housing booms and busts in between. Financial herding is an important artefact of our social nature, and the herd is a crucial conduit for speculative bubbles.

One of the most colourful historical examples of speculation was Tulipmania. For a brief period in 1637, speculators got very excited about tulip bulbs. It is not clear what triggered the excitement. There is some evidence that tulips were already fashionable, having been introduced to Europe from Turkey less than a century earlier. They were admired as an unusual and exotic flower. But interest had soon grown to such an extent that it tipped over into an extreme speculative frenzy. Traders followed each other into the tulip bulb market, chasing and initially contributing to massive speculative gains. For some of the rarer tulip bulbs, prices rose by up to 6,000 per cent. A particularly prized bulb, the exotic Semper Augustus, sold for around 1,000 florins at the height of the bubble – by various accounts more than enough money to buy either a smart townhouse, a small fleet of battleships or a

Figure 7. Jan Brueghel the Younger's tulipmaniacs: 'Satire on Tulip Mania', *c.* 1640.

drove of 3,000 pigs. The bust that marked the end of Tulipmania was as spectacular as the boom. By February 1637 most bulbs were relatively worthless, the tulip market all but disappeared. Those tulip speculators who had joined the frenzy late lost their fortunes.[10]

Tulipmania was not easily forgotten. Perhaps because it captures something essential about how our lives are driven by instinctive and unconscious motivations. In his painting 'Satire on Tulip Mania', Jan Brueghel the Younger depicted tulip traders as anthropomorphised monkeys, suggesting a primitive, basic and undesirable aspect to the speculative frenzy. Breughel's monkey metaphor speaks to something of the tensions driving our evolved instincts to follow others, unfolding in financial markets as well as our ordinary lives.

Rational bubbles

You might imagine that Tulipmania was the ultimate demonstration of an irrational speculative bubble. Certainly, from a group or macroeconomic perspective, it was destabilising and unproductive. But some economists argue that Tulipmania is entirely consistent with rational choice. They argue that rational speculative bubbles emerge as an inevitable consequence of speculators thinking carefully about the best way to make profits. For them, speculative bubbles are rational bubbles.

There is some weight to this argument. If you were a tulip trader, by observing others you might rationally judge that it makes sense to follow all those other tulip traders and buy a bulb yourself. If you had 1,000 florins, you might even contemplate buying a tulip bulb instead of a town house if you thought you could sell the bulb to the next person to join the herd for 1,100 florins. You are not being stupid if you pay an exorbitant price for something today if you think there is a good chance you can sell it to someone else for an even more exorbitant price tomorrow. The real, inherent value of that tulip bulb is irrelevant (even if you could figure out what that was).[11] The tulip traders who created the mania for tulips were just balancing the chances of the bubble persisting or bursting. For as long as the bubble was likely to continue, it was rational to spend a fortune to enter the tulip market, because that fortune might be magnified the very next day.

What drives speculators to herd together in this way? At first glance, episodes of speculative herding seem to overturn two fundamental and related assumptions that form the backbone of mainstream economic and financial theory. Economists call the first assumption the *rational expectations hypothesis.*[12] Like *Homo economicus*, which we introduced in chapter 1, economists assume that people generally, and financial traders specifically, are clever and rational. In deciding if they want to buy an asset, they must first decide what it's worth. They must form, as accurately as possible, an expectation of the asset's value in the future – if they wanted to sell it in a few years' time, for example. This expectation should reflect the *fundamental value* of the asset – what the asset would be worth if a person held on to it forever. We can illustrate the concept of fundamental value with some examples. For a homeowner, the fundamental value of a house, if they rent it out, would be all the rent it would earn its owner over its lifetime, or the rent its owner would save if they decided to live in it. For a stock or share in a company, whether listed on a stock exchange in London, New York, Riyadh or Shanghai, the fundamental value is all the dividends the stock or share would earn for as long as the company was listed on the stock exchange, and these dividends will track the profits of the listed company over time. According to mainstream financial theory, when traders form these expectations of what an asset will be worth in the future, these will track the asset's fundamental value.

In capturing the behaviour of speculators, the rational expectations hypothesis complements a second assumption from mainstream economics and finance: the *efficient markets hypothesis*.[13] This is about how the price of a financial asset – whether it is a stock, share or tulip bulb – changes over time as new information arrives. This links to the idea that, if financial markets are working efficiently, then changes in an asset's price should reflect all information, including the latest news. Share prices will fluctuate in tandem with news, good and bad, about the likely future performance of the underlying company. Fluctuations in BP's share price after the Deepwater Horizon oil spill in April 1990 illustrate the way in which share prices can change following bad news, reflecting speculators' adjusting of their expectations of future profits. Various problems with the construction of the Deepwater Horizon oil well in the Gulf of Mexico led to a blowout in the wellhead, spilling millions of barrels of oil into the ocean – with catastrophic consequences for the environment, wildlife and local businesses. As soon as news of the spill broke, speculators quickly guessed that BP's future profits were likely to be significantly eroded by compensation claims and so rapidly sold their BP shares: by June 2010, BP's share price had collapsed by over 50 per cent.

Economists also assume that speculators are acting independently of others, both in their use of information – social learning is precluded – and by looking after their own self-interest. These highly rational agents do not make systematic mistakes, and they efficiently use all the information they come across. In this sort of world, traders will trade away any difference between an asset's fundamental value and its market price. For example, if traders perceive that the fundamental value of BP shares has fallen but the market price is still relatively high, then they will sell their BP shares. Then the forces of supply and demand kick in. With lots of traders selling the shares and not many wanting to buy them, the market price will fall until it matches the fundamental value. So, profits will not persist for any length of time.

The problem with the efficient markets hypothesis and its sister rational expectations hypothesis is that they both embed extreme assumptions about markets and people. Economists know well that markets only work smoothly and fluidly when there are no market failures – but key market failures, including imperfect information and uncertainty, are endemic in financial markets. How can an ordinary person know everything they need to know about the value of the assets they buy, especially in a world plagued by uncertainty? People struggle to predict how the price of petrol might change in a day, let alone how the price of exotic, esoteric assets might fluctuate over time.

Episodes like Tulipmania illustrate that it is not as easy to be clever as mainstream economic theory suggests. That does not mean, however, that there are no good reasons to follow herds of other speculators. If you had found yourself in the middle of the Tulipmania bubble, your best strategy would have been quickly to follow other speculators into the tulip market, but make sure that you quickly followed them out of the market too. The herding heuristics introduced in chapter 3 play an important role in guiding these speculators' buying and selling choices. As we explored earlier, we use herding heuristics as a form of fast thinking. Heuristics enable us to decide quickly, without having to explore thoroughly all the potential sources of information. Instead, we employ our herding heuristics by copying what someone else is doing, assuming they have done the research already and know all that we need to know. The problem with herding heuristics in financial markets is that those markets are far from simple interactions between small numbers of people. Particularly in the modern, globalised and complex financial system, herding heuristics can trigger systemic crises that spread through financial systems and into macroeconomies more widely, as the 2007/08 US subprime mortgage crisis amply illustrates. Money is liquid and easy to trade and so errors are quickly copied and magnified. To learn more about this we can return to the theories of John Maynard Keynes.

Keynes on speculators

Keynes had a range of useful insights about speculative traders. Some foreshadowed economists' explanations for herding.[14] Others focused more on sociopsychological influences: Keynes was a pioneer in analysing the social forces driving financial markets and the macroeconomy. He focused particularly on the role of conventions in trading behaviour. In times of uncertainty, social conventions encourage speculators to believe what others believe and to do what others do.

The manifestation of this is that speculators imitate others and follow the crowd.[15] Keynes did not argue, however, that social conventions are irrational. From his early *A Treatise on Probability* of 1921 through to his major masterwork *The General Theory of Employment, Interest and Money*, Keynes' view was that conventions are a useful tool that helps us to judge the probabilities of various alternative options. In an uncertain world, expectations about asset prices are volatile because no-one knows what to expect next. Amid this confusion, the conventional opinions we share with others provide an (albeit often unstable) anchor for beliefs, calming our anxieties.[16] Keynes' speculators are chasing short-term profits and making money by quickly buying and selling financial assets. They are focused on the price they can get for the assets they are selling over the day, the week or the month – or even the millisecond, given innovations enabling high-frequency trading today. If speculators are operating in a world where they might have to sell quickly, it makes sense for them to pay very close attention to what everyone else is thinking because they may have to sell to someone else within a short period of time. So they follow conventions and scrutinise others' actions before deciding what to do themselves.

Herding and social learning

Delving deeper into his analysis, Keynes focused on three main reasons why financial investors are so preoccupied with what everyone else is doing and thinking: social learning, reputation and beauty contests. In financial markets, imitation determines whether or not we buy a financial asset and how much we are prepared to pay. We buy these assets not necessarily because we know much about their potential, but because we see others buying them and assume they know something we don't. People follow the crowd because they think that the rest of the crowd is better informed. Keynes postulated that the same process operates in financial markets. In times of uncertainty, speculators realise that they are ignorant and respond by imitating other speculators. Speculators use social information about what other speculators are buying to guide their own choices, and this tendency intensifies when information is poor and uncertainty is endemic.[17] Our decision to sell is partly driven by what we hear in the news and partly what we can see the rest of the herd doing. This links to the Bayesian social learning models of self-interested herding introduced in chapter 1. When social information overwhelms our private information, we will join a herd of copycats all choosing the same option. In this Bayesian process, speculators are using sophisticated logic. The difference in Keynes' analysis is that he focuses more on the social and psychological motivations and less on the application of mathematical tools.

There are individual differences in susceptibility to these informational influences. One example is the distinctive strategies adopted by professional as opposed to amateur speculators.[18] Amateur speculators are more inclined to imitate, but as they acquire more knowledge and private information, they become less dependent on the social signals conveyed in others' choices. Professional speculators are less likely to follow the crowd because they have a larger stock of private information and expertise. Another example is the small minority of the players in financial markets who ignore social influences almost entirely, making their money out of what seem to many other speculators like excessively risky maverick trading strategies. Famous investors George Soros and Warren Buffett, for instance, have made large fortunes from their distinctive investment strategies. So, speculators are not always copycats. Occasionally a small number of speculators may have the expertise and skills to make their fortunes from contrarian financial investment strategies.

The economist Richard Topol has constructed a general model that captures this range of speculator behaviours – from imitation driven purely by what others are doing through to the completely independent decision-making associated with the mainstream models. Topol does this by setting out a model in which speculators decide what they are prepared to pay for an asset by balancing the information they have about other traders' valuations. They have two sets of information: first, what they believe themselves is the right price for an asset, and second, the prices that other traders are willing to pay or accept when they are buying or selling. How speculators weight these different pieces of information will change depending on how confident they are in their own judgements. When copycat speculators have little confidence in their own judgements about the price of an asset, they will focus on how much other speculators are paying. They will assign a zero weight to their own beliefs. Herding will overwhelm their private judgements – in much the same way as social information overwhelms private information in the Bayesian social learning models. At the other extreme, when contrarian speculators ignore all the others then they are effectively assigning a zero weight to other speculators' prices and focusing entirely on their own judgements. Topol's model then reverts to the mainstream model in which rational, independent speculators form their judgements independently and do not worry about what the herds of speculators around them are doing.[19] In this way, Topol covers the range – from the standard economic model, based around the assumptions of rational expectations and efficient financial markets, through to the pure herding models in which speculators are completely preoccupied with what other speculators think.

Reputation

As we have already seen, preserving reputation is another reason for people to copy others. John Maynard Keynes made the astute observation that it is better to be conventionally wrong than unconventionally right. This can explain conventions in financial markets: a trader who loses £1m when his peers are also losing £1m will probably keep his job. A trader who loses £1m while others are losing nothing will almost certainly be fired.

Keynes' insight has made its way into modern economic theory, for example in the analysis of the decisions of managers of investment funds – these are the funds invested in portfolios of different financial products. The job of the investment fund managers is to convince their customers that they are investing wisely. Sometimes a fund manager will lose money because markets are inherently unpredictable and not because they made poor decisions. Then their mistakes are only mistakes from the perspective of hindsight. Given this unpredictability, fund managers will therefore rely for their reputations on comparisons with their peers, via a process of benchmarking against other analysts operating in similar markets. Benchmarking and peer comparison lead traders to focus on a different set of goals and incentives. They are being encouraged to compare themselves to others, and this leads them to follow others and disregard their own private information, even if it is more reliable.[20]

Economists David Scharfstein and Jeremy Stein use these insights to analyse herding in financial fund managers' decisions, and they explain financial herding as the outcome of reputation-building.[21] In selling their products, investment fund managers have to work hard to convince investors to invest with them. The problem is that potential investors are often more worried about short-term performance than long-term performance. But fluctuating financial markets mean short-term performance is not necessarily a good indicator of skill. Financial markets can exhibit upward momentum in asset prices in the short term, and so just because a fund manager has bought into that rising momentum it does not mean that they have a genuine and unique talent for delivering further gains in the future. Also, if their potential clients are not professionals, and are relatively ignorant, then fund managers may have no clear incentive to worry about complex performance indicators that their clients cannot understand anyway. Instead, they rely for their business on building their reputations and comparing well against their peers. In this, others' recommendations, whether disseminated via word of mouth or social media, will be a powerful influence on investment managers' ability to attract and retain their customers.

Beauty contests

Financial herding is also driven by speculators' attempts to second-guess what others are thinking. When we are deciding what we are prepared to pay for an asset, especially if we intend to sell it quickly, what other people are willing to pay for it is a good anchor for our own judgement about what we should pay. Others' willingness to pay will determine the price we might be able to achieve if we are selling the asset ourselves. Keynes described this phenomenon using the metaphor of a beauty contest.[22] He imagined a newspaper competition in which readers are asked to look at some photos of women and then judge not who they personally think is prettiest, but who they think *other readers* think is prettiest. Keynes argued that a similar process describes financial speculation: speculators buy stocks and shares at seemingly exorbitant prices not because they independently believe that these assets are really worth that much, but because they believe other speculators are prepared to pay similar prices.

Speculators' preoccupation with others' opinions has a reasonable basis. Ultimately, speculators are in the business of buying and selling assets to make a profit. They are also trading in fast-moving, highly liquid markets and they want to be able to sell very quickly, so they need to be able to match the price expectations of other traders around them. Speculators cannot afford to wait too long to find someone whose ideas about the fundamental value of an asset match their own. So, the individual speculator decides that their own convictions and judgements are largely irrelevant. For them, it is more important to know how much others are prepared to pay. How much do others think others are prepared to pay? How much do others think others think others are prepared to pay? How much do others think others think others think others are prepared to pay? And so on and so on. Keynes argued that, with everyone worrying about what everyone thinks everyone else is thinking, financial markets are not founded strongly on people's careful assessment of the likely prospects of different assets. In fast-moving financial markets, carefully assessing the facts determining the fundamental value of an asset does not help speculators to make money. Predicting what others think might.

Modern economists have adapted Keynes' metaphor in their theories of *iterated reasoning*. We form our beliefs about a collective judgement, for example about the price of a share, by iterating from one person to the next. For example: imagine I try to predict what Abu thinks a share is worth, while Abu is trying to figure out what Bob thinks it's worth. Bob is trying to figure out what Chandra thinks it's worth, and Chandra is trying to figure out what Des thinks it's worth, and so on. As for me, I have to figure out what Abu thinks Bob thinks Chandra thinks Des thinks it's worth. A lot of cognitive effort is required to figure out what the crowd, as a whole, thinks about the value of a share. We might judge (sensibly) that it's not worth making all that cognitive effort when we can just copy the next person by paying what they pay. More importantly, if no-one else is thinking very deeply about the problem, then it is pointless for us to think deeply about it. We will do much better if we just copy the herd.

Experiments based on beauty contests in financial settings have confirmed that many people are not very good at reasoning far into these iterative thinking problems. Some of these experiments analysed decisions by CEOs and other readers of the *Financial Times*, audiences we might expect to have a relatively sophisticated knowledge of finance. Even the CEOs did not reason deeply about the beauty contest game.[23] For those who did try to reason through the example given above, most of them got as far as worrying about what Des was thinking, and then stopped trying to second-guess any further. Their failure to think beyond Des was not necessarily because they were not capable of reasoning more deeply. They may have made the strategic choice not to think too deeply because they guessed that others wouldn't get very far with it either. Their best guess just needs to match the next person's.

The problem is that a world in which everyone is worrying about what everyone else is thinking is a breeding ground for financial instability, and this was one of Keynes' fundamental points. It is important to emphasise that this is not a stupid strategy for each individual speculator. If a speculator just wants to make money quickly, then it makes sense for them to focus on what the herd is doing and paying – from their own perspective at least. From a collective, social or macroeconomic perspective, however, when this preoccupation with what others think is aggregated across many individuals interacting within complex financial systems, financial markets transform into incubators for financial disaster. No single individual has any incentive to figure out what assets are really likely to generate in real terms in anything beyond the very near future. If no-one is worrying what an asset is likely to deliver in real terms, then there is no guarantee that money will flow towards the most productive and efficient businesses and projects. As Keynes observed:

> Speculators may do no harm as bubbles on a steady stream of enterprise. But the position is serious when enterprise becomes the bubble on a whirlpool of speculation. When the capital development of a country becomes a by-product of the activities of a casino, the job is likely to be ill-done.[24]

The preoccupation with others' opinions and conventions destabilises financial markets. When the price we are willing to pay for a financial asset is so far removed from our own personal judgement of the fundamental value of an asset, then the herd's judgement overall becomes flimsy and unstable. Instability is magnified particularly with short-termist, impatient speculators who want to buy then sell as fast as they can to make a quick profit.

Emotional herding

So far, we have focused on economic explanations for speculators' susceptibility to social influences. Individual differences, especially personality traits, will play a role in determining whether the contrarian or copycat side dominates. As we saw above, the social learning model suggests that the balance of private and social information will determine whether a speculator is more or less likely to follow the crowd, and the well-informed professional speculators are more likely to adopt a contrarian strategy. More subjective factors will drive financial herding too, including psychological and emotional influences. For example, impulsivity is an important trigger for herding, and may connect with evolved instincts, if following the herd is an automated, instinctive response. There are also possible connections with other personality traits associated with sociability. Psychological measures of conformity and extraversion are very likely to correlate with financial traders' propensity to follow the herd, though the extent of this correlation will depend on whether a financial trader is an amateur or a professional. Personality traits will also determine a trader's susceptibility to emotional influences. Emotions play an important role in our financial decision-making, especially as many financial decisions involve risk-taking, which is often emotionally charged. Financial analysts are increasingly acknowledging the impact of these biological, innate and instinctive responses to stimuli on their working lives, particularly in the context of basic emotions such as greed, hope and fear.[25]

External events also have an impact. Even the weather can play a part. Some economic researchers claim that financial performance is affected by seasonal mood changes: for example, Mark Kamstra and colleagues have shown that trading performance is impaired during wintertime, and attribute it to seasonal affective disorder.[26] David Hirshleifer and Tyler Shumway have shown that stock market patterns around the world are correlated with hours of sunlight.[27] Researchers at the Socionomics Foundation based in Gainesville, Georgia have suggested that all economic and financial instability, including financial herding, can be explained by fluctuations in social mood. Maybe this is not so surprising: social mood impacts on all aspects of our lives. Trends in music, fashion, construction and literature are all propelled by social mood.[28]

Bringing these insights together, social emotions, propelled by shifting moods across markets and economies, drive herding in financial markets.

Financial herding: cognition, emotion and neuroscience

In the case of Tulipmania, were the tulip traders caught up in one of Keynes' beauty contests and rationally paying high prices because they thought someone else was likely to pay even more the next moment? Or were they getting carried away with the excitement of it all, driven by some fast-thinking, emotional buzz akin to addiction? Economists have disagreed over the extent to which speculative frenzies such as Tulipmania are rational or emotional.[29] In reconciling the apparent contradiction, we can return to Kahneman's dyad of System 1 fast thinking and System 2 slow thinking, and the division of effort between the two – introduced in chapter 3. If we agree that decisions are driven by more than one decision-making system, then the economist's traditional distinction between what is rational and what is irrational becomes redundant. Speculation is neither rational nor irrational. It is more likely to be the outcome of complex interactions between System 1 and System 2.

In fact, the idea that economic and financial thinking might reflect an interplay of different thinking systems was anticipated by Keynes. He captured how reason and emotion interact, in a battle between our rational and our whimsical, sentimental selves:

> We should not conclude from this that everything depends on waves of irrational psychology. On the contrary, [our confidence about the future] is often steady, and, even when it is not, the other factors exert their compensating effects. We are merely reminding ourselves

> that human decisions affecting the future, whether personal or political or economic, cannot depend on strict mathematical expectation, since the basis for making such calculations does not exist; and that it is our innate urge to activity which makes the wheels go round, our rational selves choosing between the alternatives as best we are able, calculating where we can, but often falling back for our motive on whim or sentiment or chance.[30]

How can we measure these interacting thinking styles to analyse the links between emotions and financial herding? As we noted in chapter 3, with conventional economic analysis, data about people's observed choices is relatively easy to collect. There are many large databases around the world showing the volumes of assets traded and the prices paid for them. Yet, although they record actual decisions, these databases cannot record the interactions of cognition and emotion that drove the decisions. Capturing these underlying influences on financial traders' decisions is becoming easier as neuroscientific techniques improve.

As introduced in chapter 3, neuroscientists link financial decision-making with the neuroscientific evidence by showing that money stimulates the same neural reward-processing systems activated by the pursuit of rewards such as food, sex and drugs. In one study, researchers monitored professional derivatives traders' physiological responses while they were engaged in risky gambles. The traders experienced heightened emotional states, measured in terms of elevated heart rates, muscular responses, high blood pressure, rapid respiration rates and elevated body temperature. Experienced traders were generally better at controlling their emotions.[31] In another study, researchers examined people with brain damage in specific neural areas including those usually associated with emotional processing, such as the amygdala and insula. People with damage to their neural emotional processing circuits were more willing to take risks by investing money in gambling tasks. They also made larger profits than the experimental subjects in a control group, perhaps because decreased affect ameliorates problems created by more impulsive decision-making. We explored above why speculators may be inclined towards myopia and short-termism in their buying and selling decisions. They are excessively preoccupied with ephemeral, day-to-day fluctuations. This interacts with their fear of *losing* money through their trading activities – reflecting the phenomena of loss aversion explored in earlier chapters. Nobel Prize-winning behavioural economist Richard Thaler, working with his colleague Shlomo Benartzi, brought together insights about myopia and loss aversion by identifying a financial decision-making anomaly: *myopic loss aversion*. Myopic loss aversion is a bias that emerges when speculators are simultaneously too focused on the short term and excessively preoccupied with losing money. It distorts the balance between risky equities (e.g. shares in companies) and safe bonds (e.g. bonds representing a piece of government or corporate debt). Why is it so distorting? If financial markets are working well, we would expect speculators to buy into assets that have higher returns, but, because of myopic loss aversion, speculators worry excessively about losing money quickly if they buy equities and so they buy fewer equities than they need to maximise their profits. Instead, they are disproportionately inclined to buy bonds, even though returns on bonds are lower. The differences in returns on equities versus bonds are not traded away, and traders do not maximise their profits.[32]

The social influences we have explored in this chapter increase the intensity of speculators' emotional responses, and this can be seen in financial markets when social conventions encourage speculators to believe what others believe and to do what others do. Emotions are processed much more quickly and easily than quantitative and mathematical information, and they spread more quickly through the herd, magnifying financial instability. Drawing on similar insights, some economists describe phases of boom and bust as manic-depressive episodes, driven by emotions. As American economist Hyman Minsky observed in the 1980s and 1990s (well before the financial instability of 2007/08), during an economic boom, speculative euphoria spreads quickly through entrepreneurs, investors and bankers, catalysing surges in construction activity and financial bubbles. But because the bubble is unstable, it can quickly burst. Individuals panic, and their panic spreads. As negative unstable forces take hold, economies and financial systems lurch into crisis, with excessive pessimism and extreme risk aversion precipitating bust phases. As we explore in more depth below, Minksy's analysis predicted that recession and depression would emerge in the aftermath of a perfect social storm of risk, anxiety and fear.[33] Minsky's analysis chimes with recent evidence from psychological studies suggesting that interactions between risk, emotions and herding intensify fearfulness and trigger social panics. Panicking individuals precipitate panic through the herd.[34]

Entrepreneurial mavericks

In the previous chapter we discussed different types of mavericks – people who are prepared to take risks with new and different ideas. Economies are driven by two specific types of mavericks: inventors and entrepreneurs. Inventors are a classic type of maverick and their novel inventions are fed into the economy via another set of mavericks: entrepreneurs. Entrepreneurs are prepared to take risks in turning an invention into an innovation and then into a marketable product or service. The renowned economist Joseph Schumpeter captured something of how herding and imitation drive innovation and entrepreneurship in the economy. For Schumpeter, innovative entrepreneurs are heroes. They are the lifeblood of a successful capitalist economy and, when they introduce a new business idea, they attract swarms of imitators who want to copy them. At the outset, many of these imitators will benefit from the profits the new innovation brings, but eventually, when the swarm of copycats grows too large, the benefits will disappear, and the economy as a whole will head into a downturn.[35]

Another famous account of maverick entrepreneurship comes from John Maynard Keynes in *The General Theory of Employment, Interest and Money*:

> it is the long-term investor, he who most promotes the public interest, who will in practice come in for most criticism . . . For it is in the essence of his behaviour that he should be eccentric, unconventional and rash in the eyes of average opinion. If he is successful, that will only confirm the general belief in his rashness; and if . . . he is unsuccessful . . . he will not receive much mercy . . .[36]

Keynes also emphasised the far-sighted nature of entrepreneurship. Ephemeral influences will not help entrepreneurs to make good decisions, especially as the rewards from good business ideas are unlikely to emerge over short time horizons. Keynes observed that an entrepreneur is unlikely to be able to calculate the future prospects of their business projects because the future is inherently uncertain – and so entrepreneurs need to be forward-looking and optimistic. Entrepreneurs realise that it takes time to generate profits and so have a patience that financial speculators often seem to lack, especially in new, innovative industries. Facebook, Instagram and Twitter are all examples of innovative businesses which did not immediately deliver revenues and profits, yet their founders had a vision of what their companies could become in the future. During the dot-com boom of the 1990s many businesses failed – and perhaps their founders were also forward-looking mavericks, just unluckier or with an inferior product.

Uncertainty about the future constrains effective decision-making by maverick entrepreneurs, but they are less susceptible to herding than most speculators. Building a business is not usually about sitting down with the accounts and making an arithmetic calculation of likely future profits, partly because it is difficult to predict the future and the information needed to make such calculations just does not exist. Entrepreneurs are not looking to make money out of short-term fluctuations in fast-moving markets. Social learning, reputation, beauty contests: all these factors have a lesser impact on entrepreneurs than on speculators. Entrepreneurs look to the long term, and so the short-termist opinions of others around them are not so relevant. Overall, the contrarian entrepreneur is less vulnerable to herding's negative impacts than the consensual speculator. Instead, entrepreneurs rely on their internal intrinsic motivations, and they take an optimistic view of what might happen.

Social influences are not irrelevant to entrepreneurs, but they affect them in different ways. Daron Acemoğlu explored social information from the perspective of entrepreneurial investors in his model of *signal extraction*. Entrepreneurs extract signals from macroeconomic data, for example data on fixed asset investment – the money spent on things like machinery and buildings – making inferences about what other entrepreneurs are deciding using this aggregate information. This helps each individual business person to judge a situation, such as the wisdom of investing in a new business. In a macroeconomic corollary of the self-interested herding models we explored earlier, aggregate information helps individual entrepreneurs to infer what other entrepreneurs are doing.[37] By looking at aggregate data about what everyone is doing collectively, entrepreneurs can extract signals about likely future prospects of their new business ventures.

Entrepreneurial emotions

Entrepreneurs' far-sightedness does not mean that they are immune from psychological influences. A recent study into small businesses in Africa has shown that psychological traits associated with initiative-taking and goal-setting are associated with better business performance than traditional business education.[38] Another feature of the entrepreneurial personality is that entrepreneurs are likely to be people of action with a strong urge to do things differently – a reflection of their contrarian natures. When they bring new innovations to the marketplace, entrepreneurs are motivated not only by the profits they might earn, but also from the psychological satisfaction they get from building a business. They have stronger maverick inclinations and are more likely to be propelled by gut feeling and other emotional and psychological influences into getting something done. Keynes describes entrepreneurial mavericks thus:

> Most, probably, of our decisions to do something positive, the full consequences of which will be drawn out over many days to come, can only be taken as the result of animal spirits – a spontaneous urge to action rather than inaction . . .[39]

This concept of 'animal spirits' links back to the ancient Greek physician Galen, first introduced in chapter 3, and his analysis of four temperaments. In developing his concept of animal spirits, Galen followed in the footsteps of Hippocrates, another renowned ancient physician and philosopher who postulated that our behaviour is driven by four 'humours', each of which was linked to four essential elements: black bile to earth, blood to air, phlegm to water and yellow bile to fire. Galen developed Hippocrates' schema by linking each of these humours to a different temperament: black bile is melancholic, blood is sanguine, phlegm is phlegmatic and yellow bile is choleric.[40] Related to these humours, Galen popularised the idea of 'animal spirits'. These are something like a sub-category of neurotransmitters, the chemical messengers that flow around our body, through the nerves, and help its functioning. For Keynes, 'animal spirits' were a way of conceptualising entrepreneurs' sanguine temperament. He observed that 'investment depended on a sufficient supply of individuals of sanguine temperament and constructive impulse who embarked on business as a way of life'.[41] Whilst Galen's ideas seem naïve from a modern medical perspective, Keynes' saw animal spirits as a means of explaining the positive attitude of entrepreneurs towards innovation as investment, now a focus of modern models of behavioural macroeconomics, as we shall see.

Ecology: copycat–contrarian symbiotics

We can see easily that entrepreneurs are valuable players in our economy. They produce things. They employ people. They don't worry what everyone else thinks. The importance of speculators to our economy is less obvious because they do not produce anything physical of value themselves. So, why do we need them? They are the inevitable product of the financial markets on which entrepreneurs depend. Financial liquidity is important for any entrepreneur looking to build or sustain a business venture, and fast-moving financial markets can help entrepreneurs to raise money quickly for new investments. Before the advent of modern financial markets, if an entrepreneur wanted to invest in a new business they would have had to either raise funds from their own resources or go to a bank. With stock markets, they can access finance much more quickly. Entrepreneurs need financial markets and financial markets need speculators to keep the money moving around. For this reason, entrepreneurs and speculators have developed a symbiotic relationship.

The link between the speculators' activities in financial markets and the needs of entrepreneurs is explained by Keynes in his *General Theory*. Keynes argues that there is no sense, at least in terms of easy ways to make money, in building up a new business if you can find and invest in the same enterprise on the stock market by buying its shares. So, there should be a link between the market valuations of companies listed on stock markets and entrepreneurs' incentives to invest in building up businesses. Individual speculators and entrepreneurs would find it difficult unilaterally to coordinate their supply and demand for funds to build a new business, hence the need for the financial market.

What consequences do speculators' actions have for entrepreneurial activity? In their book *Animal Spirits*, George Akerlof and Robert Shiller develop the connections between emotions and Keynes' concept of animal spirits to explain how economic and sociopsychological factors feed off each other in the interplays between entrepreneurship and speculation. Akerlof and Shiller define animal spirits more broadly than Keynes – not only as psychological influences driving entrepreneurs, but as including a range of different psychological influences distorting the economy and financial markets.[42] For Akerlof and Shiller, a particularly powerful psychological driver is storytelling. They argue that social storytelling helps to explain financial herding across different types of markets, for example in housing markets. In homebuyers' minds, they join the herd in buying into a housing bubble in the false belief that house prices can never fall.

They believe this because the dominant, conventional story told through the pronouncements of politicians and policy-makers, news stories and word-of-mouth information is that house prices only ever rise. Akerlof and Shiller argue that in this way, naïve stories and folk wisdom fuelled the excessive increases in house prices across global markets in the 1990s and 2000s.[43] Misguided consensual opinions allowed the bubble to grow too fast, magnifying the consequences of the collapse when it finally came. All this was exacerbated by the perverse incentives to buy into housing assets, especially when the large bonuses that financial traders could earn from these assets were added to the mix. Herding was not only driven by the interactions between copycat speculators, it was also enabled by the other actors and financial market institutions that did not challenge the flimsy foundations on which trading activities were based – including financial market regulators, credit rating agencies, politicians, academic economists – even journalists.[44]

These ideas return us to the insights of Hyman Minsky introduced earlier in the chapter. Minsky also analysed interactions between entrepreneurship and speculation across economies. If people and businesses are feeling optimistic, then a kind of euphoria will take over the macroeconomy: entrepreneurs will want to invest in their businesses, and perhaps build new ones. Bankers will be keen to lend plenty of money, and at lower interest rates. Speculators will thrive in this environment. As a boom begins, demand for plant, equipment, factories and housing will expand, and the construction sector will thrive on this growing desire for new and bigger buildings and infrastructure projects. However, the economic and financial system cannot continue on this upward trajectory forever. Soon, tensions and cracks will emerge, as people, businesses and banks start to realise that the levels of debt incurred during the boom phase are not sustainable.

Overconfidence and optimism will be replaced by underconfidence and pessimism, and, in a mirror image of the boom phase, word will spread through the herd that prospects are not so good after all. Storytelling, word of mouth and false intuitions go into reverse, feeding herding and contagion as asset prices fall.[45] The 2007/08 global financial crisis and the collapse in subprime mortgage lending that preceded it is a powerful example of the destructive power wielded by narratives and stories when they distort the delicate balance between entrepreneurship and speculation in the economic-financial ecosystem – with wide-ranging impacts on entrepreneurship, production and employment more generally.

Controlling speculation, encouraging entrepreneurship

We have seen the ways in which copycat speculators can have profound and destabilising effects on modern economies. Entrepreneurs are more generally the heroes of the economy. When financial markets are computerised, globalised and overconnected then financial contagion spreads rapidly. Something like the butterfly effect that characterises chaotic systems will take hold, as explored by British economist Paul Ormerod.[46] Small groups of speculators introduce large amounts of instability – not just into the financial system but also into economic production and employment. One illustration of how the actions of lone individuals can be magnified spectacularly across the globe is the 2015 flash crash on the Chicago Mercantile Exchange. Navinder Singh Sarao was convicted in 2017 for manipulating financial markets for profit via the practice of 'spoof-trading'. He executed large, false 'sell' orders to push down share prices, triggering herds of speculators to sell because they saw others selling. Sarao then bought back the shares at the new, lower prices, cancelling his initial orders to reverse the share price falls. He made a lot of money from buying at the low price he had himself engineered and then selling as prices rose again when he reversed his spoof trade. And all this was done from a computer in the bedroom of his parents' home in Hounslow, west London. In globalised, computerised and deeply interconnected financial markets, one person can engineer enormous financial damage by manipulating herds of speculators.

From the perspective of evolutionary biology, there is no reason that an ingrained instinct to herd should be useful in modern financial markets. If financial speculation is little more than a form of institutionalised gambling, then perhaps our primitive fast-thinking instincts are not well suited to the modern world of globalised, computerised financial systems and sophisticated modern innovations such as algorithmic trading. Does herding driven by fast thinking magnify the fragility of the financial system? Yes, because if traders' rewards are determined by short-term performance then all the speculators' tangible incentives and unconscious instincts line up to support impulsivity. If large-scale herding in financial markets reflects the overriding influence of hardwired emotional responses and generates excessive financial instability, then it may be a maladaptation to be discouraged.

In essence, herding has benefits as well as drawbacks, and that is true in the economy too. Some speculation is a good thing, because it funnels financial resources towards business investment, entrepreneurship and employment. There is wisdom in crowds and collective opinion can under certain specific circumstances be more accurate than individuals' opinions. Condorcet's wisdom of crowds hypothesis, which we introduced in chapter 2, assumes that individuals form their judgements rationally and objectively, and allows no role for psychological and emotional influences. Yet the assumption of independence highlights one of the problems to do with financial herding and financial instability. Individual traders' judgements are not independent. Different traders feed off each other's decisions, and they use the same information to arrive at their judgements. There can be severely negative impacts for entrepreneurs and economies more generally. Either way, policy controls are needed to keep financial herding in check so that its benefits can be realised and its downsides contained.

How – and by how much – should governments intervene to moderate financial speculation and/or encourage people towards entrepreneurship? In the aftermath of the 2007/08 financial crisis, this is the major question that politicians, governments and international institutions have yet to answer. Once we are better able to understand what triggers financial herding, then we will be better able to control financial instability and its wide-reaching economic consequences. Some policy instruments have been suggested to slow financial markets. Impacts on the wider economy are being reduced in some countries by disentangling retail banking for ordinary customers from investment banking, the most risky and unstable part of modern financial markets. If more widely adopted, this will limit the impact of volatile investment markets and rising interest rates on consumers and entrepreneurs, as well as increasing the volumes of lending available to the private sector. A tax on financial transactions – the so-called 'Tobin tax' – could put sand in the wheels of fast-moving markets driven by excessive herding and other destabilising influences. The problem, though, is that to be effective this sort of tax would have to be adopted globally, to prevent capital flight to tax havens.

Assuming that herding with the market is an impulsive, emotional response, if a way could be found to turn traders 'off' when their decisions are becoming too emotionally charged, then this could reduce destabilising speculative activities. A more powerful solution would be to rethink corporate governance and institutional arrangements. Financial services companies have instituted policies to ensure more effective monitoring of their traders' activities, including more careful line management and team structures that ensure that the activities of individual traders are more controlled and less emotional, limiting 'rogue' traders' room for manoeuvre. Policymakers are also recognising that herding and social influences destabilise financial markets and economies. They are exploring ways to resolve the problems. In the UK, the Financial Conduct Authority is developing research and insights to understand and ameliorate problems of herding and groupthink in asset management, especially by institutional investors.[47] Similar initiatives have been conducted in the US by the Securities and Exchange Commission. If the oversight committees for large institutional funds can be more effectively designed to reduce the influence of unsubstantiated consensus, then this will be a check on the herding behaviours that destabilise financial markets and entrepreneurship.

Examining the practices of speculators and entrepreneurs has introduced us to the idea that our economies and financial markets are characterised by complex interactions between copycats and contrarians. Speculators are more usually copycats. Entrepreneurs are usually more contrarian. Sometimes, however, a speculator might do well to take maverick risks and magnify their returns in consequence. Entrepreneurs might do well to imitate the innovations of others in developing their new business models and strategies. But economies are not the only ecosystems in which copycats and contrarians come together. They do so in many other areas of our lives too – for example in scientific research, politics and religion, as we shall explore in the next two chapters.

7

Herding experts

In 2009, Trevor Ulrich, a toddler from Maryland, died suddenly. His post-mortem revealed bleeding on the brain and scalp contusions. His day-care minder, Gail Dobson, was accused of his murder, though there were no witnesses to the supposed crime. At the trial, following the conventional medical opinion of the time, the court-appointed doctors testified that Trevor had died of what was then known as 'shaken baby syndrome' – from injuries sustained through being violently shaken, perhaps in a fit of anger. The final verdict has gone back and forth, but at the time of writing Dobson had eventually been found guilty of child abuse and murder.[1]

This verdict was founded on a medical diagnosis first outlined in 1974 by paediatrician John Caffey. When a baby is shaken violently, a triad of medical symptoms ensues: swelling in the brain, bleeding around the skull and bleeding around the retina at the back of the eyes.[2] Trevor suffered all of these symptoms. Following the trial, however, medical opinion started to shift. Since 2001, many shaken baby syndrome convictions have been overturned in the US. In

2009 the American Academy of Pediatrics acknowledged that the causes of the triad of symptoms are not well understood and recommended that doctors stop using the term 'shaken baby syndrome' and use 'abusive head trauma' instead. A judge's summation after he had quashed the conviction of a mother of two serving a long prison sentence included his opinion that shaken baby syndrome was 'more an article of faith than a proposition of science'.[3] The fact that innocent people were wrongly convicted does not mean that a lot of the successful prosecutions were unjust. It does suggest, however, that courts need to be properly cautious in interpreting the evidence.

Dr Waney Squier, a medical doctor, was caught up in parallel controversies in the UK. In her early days as an expert witness for British legal teams prosecuting child abuse defendants, she had embraced the consensus view about the triad of symptoms being linked to shaken baby syndrome. But over time she changed her mind, coming round to the contrarian minority view that the triad of injuries can be caused by babies and toddlers injuring themselves in other ways. Her revised expert opinions were controversial. In 2010, the National Policing Improvement Agency complained to the UK's General Medical Council that Squier's expert opinions were biased by her subjective distortions of the scientific evidence. She appeared before the Medical Practitioners Tribunal Service, which in March 2016 concluded that she had misrepresented the evidence and overreached her brief as an expert witness.[4]

Many of Squier's colleagues and friends raised concerns that she had been unjustly silenced.[5] A group of lawyers and scientists defended her in a letter to the *Guardian* in 2016:

> Every generation has its quasi-religious orthodoxies, and if there is one certainty in history it is that many beliefs

> that were firmly held yesterday will become the object of knowing ridicule tomorrow ... However, the case of Dr Squier follows another troubling pattern where the authorities inflict harsh punishment on those who fail to toe the establishment line ... It is a sad day for science when a 21st-century inquisition denies one doctor the freedom to question 'mainstream' beliefs.[6]

Squier can no longer be called as an expert witness in court trials. As a consequence, some expert witnesses are now reluctant to testify in court. Dr Irene Scheimberg, another doctor sceptical about shaken baby syndrome, told the BBC's *Newsnight* programme that she no longer provides evidence because she is 'afraid of the possible consequences'.[7]

Dr Squier took what might seem in retrospect to have been extreme risks in expressing a strongly held opinion. She has paid a very high price for her minority view, even though that view is shared by a significant number of experts. It is perhaps too soon to tell if she, and they, are right or wrong. The General Medical Council did not see it in these terms, arguing instead that Squier had manipulated and distorted the facts. But was it also her error of judgement to argue so publicly against the consensus view on a question that, objectively, remains unanswered and is still a matter of heated debate amongst experts? As an economist, I have no way of judging the scientific merits of either side, but Dr Squier is not alone in being ostracised for holding a contrary expert opinion. Whether or not she is vindicated we shall only discover in time.

Happily, time can turn the tide for a scientist prepared to take on the consensus. The Australian medical scientist Barry Marshall is famous for his self-experimentation with the life-threatening bacterium *Helicobacter pylori* (or *H. pylori*). He and his colleague Robin Warren suspected that *H. pylori* was implicated in the development of stomach ulcers, but there was no easy way to test their hypothesis. Deliberately infecting people with the bacterium would be unethical. At first, their hypotheses were ridiculed. The then prevailing consensus view was that stomach ulcers are the product of poor diet, hyperacidity and stress.[8] Marshall – bravely or recklessly, depending on your perspective – drank a life-threatening concoction containing *H. pylori* himself. He quickly developed gastric symptoms, but after prescribing antibiotics for himself, made a full recovery.[9] The experiment was a success: Marshall had been able to prove, insofar as proof is ever possible, the links between the causes and effects of stomach ulcers, as well as the cure.[10] It is now established expert opinion that *H. pylori* is the culprit in the pathogenesis of stomach ulcers, and antibiotics are now the best available treatment for them.[11] This was one of the twentieth century's most important medical breakthroughs, given that around 2 per cent of people suffering from stomach ulcers go on to develop stomach cancer.[12] On some estimates, Marshall and Warren's findings have saved hundreds of thousands of lives, and the two men were justly awarded the Nobel Prize in Physiology or Medicine in 2005.

Scepticism about contrary expert opinion is not a new phenomenon. In 1633, Galileo Galilei was convicted of heresy for arguing the truth of Copernicus's heliocentric astronomical model – in which the Earth and other planets revolve around the Sun, in contrast to the older Ptolemaic model in which the Sun and other planets revolve around the Earth. We would seem very foolish today if we argued against what is now the Copernican consensus, just as we would seem ridiculous were we to claim that the Earth is flat. Science has resolved these questions, but only after heated and sometimes violent debates.[13] Galileo and the other cases above illustrate that contrarian experts are often the targets of the most vituperative attacks. What might lead an expert herd down the wrong research track? What might motivate a contrarian expert to move against the herd's consensus? A complex set of social and individual influences, incentives and motivations are interacting, fuelled by the problem that it is rarely easy to identify an incontrovertible truth.

Fallible experts

Our everyday lives are saturated with expert opinions and judgements. Expert journalists and media pundits tell us how to interpret the latest political and economic news. Expert doctors diagnose our symptoms and prescribe treatment. Expert mechanics check our cars and boilers and tell us when they need replacing. Expert hairdressers convince us to try a new hairstyle. Expert weather forecasters advise us whether or not to travel or take an umbrella. For better or worse, expert opinions can have an enormous impact on us: consider the potentially life-changing epidemiological expertise that controls which pharmaceuticals and vaccines we do or don't use, for instance, or, as we saw above, the serious consequences arising from evidence given by expert witnesses in legal trials. Sometimes experts are right. Sometimes they are wrong. Either way, we can't know, or even assume, that their expert opinions are unbiased, driven primarily by well-informed and robust assessments of objective evidence.

We live in an age when many are sceptical about experts and their opinions. Experts are often vulnerable to fierce, critical attack – especially as modern news is so dominated by unreliable tabloid journalism and social media. But our disenchantment with experts is not new. Nor is it necessarily a bad thing. Medical quackery,[14] once favouring lobotomies and cold-water cures, now encompasses unorthodox medical, surgical and nutritional fads. With the right celebrity endorsement, the consequence can be iatroepidemics – epidemics of treatment-caused diseases – in which an element of faith is strong. Social influences have effects when they lead people to adopt a treatment just because we trust others who are advocating it.[15] Almost by definition, if we are amateurs, we should sometimes have faith in experts.

There is plenty of evidence that our attitudes and responses towards expert opinions are at best confused, and this confusion is often magnified by sloppy standards of journalistic reporting of science 'news'. We want opposing things from our experts: we want them to be original and innovative, but at the same time we are reassured by high levels of expert agreement. Forecasters of everything from the economy to the weather are often vilified for deviating from a common judgement – and they are judged only with the benefit of hindsight.

Michael Gove MP is famous in some circles and infamous in others for his opinion that 'People have had enough of experts'. He voiced his words in a Sky News broadcast in the run-up to the UK's 2016 EU referendum, during which he refuted the opinions expressed by most economists, including experts from the very reputable Office for National Statistics and the Institute for Fiscal Studies, that leaving the EU would have serious negative economic consequences for the UK. Gove's quote was perhaps unfairly truncated from his original statement: 'The people of this country have had enough of experts from organisations with acronyms, saying that they know what is best, and getting it consistently wrong'.[16] Nonetheless, his anti-expert views were clear enough, in this interview and his subsequent broadcasting, social media and print media appearances. In the build-up to the June vote, Michael Deacon, the parliamentary sketch-writer for the *Daily Telegraph*, published a clever satirical piece that put the specious nature of Gove's arguments in sharp focus. Do we really need doctors? Or pilots? Or maths teachers?

> The mathematical establishment have done very nicely, thank you, out of the notion that 2 + 2 = 4. Dare to suggest that 2 + 2 = 5, and you'll be instantly shouted down. The level of groupthink in the arithmetical community is really quite disturbing. The ordinary pupils of Britain, quite frankly, are tired of this kind of mathematical correctness.[17]

Gove's comment is easy to satirise, but that does not mean that he was wholly wrong. Scepticism about experts has been growing, not helped by the contradictory précis of experts' health and lifestyle advice we read all the time in the popular press.[18]

Experts do not help themselves though, with their poor communication styles. The public does not realise, and perhaps is not encouraged by modern media to realise, that experts are not astrologers. Experts do not claim perfect foresight. They are forming their judgements in an uncertain world in which the future is unknown and sometimes unknowable. Sometimes we forget that most of the time we want an expert's opinion on something because no-one knows the truth. The evidence is uncertain and unclear. The essence of very uncertain phenomena is that good data are scarce and difficult to interpret. It can be hard to predict future trends in complex phenomena – be they storms, stock market fluctuations, oil reserves or the spread of epidemics. We turn to experts for an answer, forgetting that experts are fallible humans and sometimes have no reliable ways of identifying the truth or forecasting future events. In an uncertain world, experts themselves are unsure, and should admit that they are unsure.

Given all this uncertainty, experts present us with the *chances* of one event or another. Their predictions are not much more than informed probabilistic guesses. Scientific experts properly acknowledge the chance element in their predictions – they would not get published in any reputable journals if they did not. But these caveats are often lost in the translation to popular media, especially social media, where experts' research and judgements are condensed into tweets of 280 characters or fewer.

Another bias is that we don't always give expert opinions the extra weighting they deserve given that experts are people with a deep, specialist knowledge of their subject. The best of broadcasting organisations, including the BBC, have been criticised for giving equal time to both amateur and expert opinions, on the implicit assumption that both amateurs and experts are equally well-informed. Is the implication that years of education and research count for nothing because everyone's opinions should be weighted equally? This trend was particularly controversial in debates between scientists and climate-change deniers. In 2014, the BBC Trust undertook a review that reiterated that not all different opinions are equal: scientific evidence and experts' opinions should be weighted more strongly than those of amateurs not grounded in a comprehensive knowledge of a subject.[19] Modern technologies may be to blame for our disenchantment with experts because they enable quick dissemination of unsubstantiated opinion as if it were fact. The consequences are that when scientific research falls into disrepute, funding trickles away too.[20] So, we need to understand better where the pitfalls lie. When experts present information to us in an authoritative way, using esoteric and technical language, we need to remember that they too are susceptible to herding and social influences. These influences might affect experts consciously or unconsciously, and sometimes malignly.

If you were to ask an expert what their goal is, they would (hopefully) answer that it is to find some objective truth via a balanced assessment of existing evidence. An academic expert would add that they aim to develop the existing research and uncover facts, following a robust and balanced scientific method. All of this pretends that experts are essentially machine-like information processors. We expect them to absorb some data, process it and churn out the best objective judgement they can. If their judgement is wrong, then we conclude that they must be mad, bad or stupid – or maybe some combination of the three. We forget that experts are social animals, just like the rest of us.

Sociable experts in an uncertain world

Social influences have more traction in an uncertain world. How do we unravel all these influences in assessing experts, given that often we have no absolute, objective benchmark of truth against which we can judge the quality of an expert's opinion? As we have seen in previous chapters, people are more likely to follow a crowd if their own information is muddy. Subjective social influences have more traction when the objective truth is very hard to find. In an uncertain world, experts do not deliver facts, they interpret data. For an economist to predict what might happen in the next year to house prices, oil prices or government deficits is really an enormous task. In these situations, the expert opinion is often just that – an opinion, not a statement of fact. Housing markets, for instance, are driven by so many unpredictable and complex factors that it is not surprising that economic forecasts have such a bad reputation for unreliability. Admitting they are unsure is sometimes the expert's most honest answer, and their best course of action is to collect more information so that the uncertainty diminishes. The problem is that woolly answers about what an expert does not know are not newsworthy. People do not want to hear that even an expert cannot really be that sure.[21]

What has this to do with copycats and contrarians? We have seen that when information is fuzzy and facts are unclear, social influences can be strongest. Herding takes a powerful hold over opinion, judgement and belief. It magnifies the difficulties inherent in interpreting complex data and evidence. In general, the evolution of knowledge is a social process. Learning about others' research happens in social contexts – at conferences, symposia and seminars. Research is mostly collaborative, and good research builds on what has gone before. As Isaac Newton observed, borrowing a metaphor attributed to the French philosopher Bernard of Chartres, 'If I have seen further, it is by standing on the shoulders of giants.' Given a leg-up by the pioneering thinkers who came before us, we can see further and understand better. And, given certain assumptions, a collective judgement may be more accurate than that of any one individual.[22] As captured by Condorcet's wisdom of crowds postulate, introduced in chapter 2, if many experts pool their beliefs, then the collective knowledge outcome may be more powerful than one expert's opinion alone – but only if the individuals' beliefs start off as independent and uncorrelated. Contrary evidence may be richer and more informative than evidence that just confirms what we already know. Contrarians play an important role in discovering new, surprising knowledge and upsetting the herd consensus. Beryl Lieff Benderly has observed that new ideas are not always welcomed within the scientific culture.[23] So the novel ideas on which progress depends do not always and easily find their way into the light.

We can look at the problems from two complementary perspectives, reflecting the underlying theme in this book: self-interested herding, which can be explained using economic theory, and collective herding, driven by sociopsychological influences. The former involves the expert's promotion of their own individual self-interest, and can be explored through the incentives that motivate and mould the individual experts' pronouncements. The latter is more complex, particularly in terms of quantifying its impacts.

Self-interested experts

In judging expert opinion, we are trying to disentangle not just whether experts are right or wrong, but also what motivates them to disagree with one another. Is it a genuine opinion, based on the interpretation of solid evidence? Are contrarian experts mendacious curmudgeons, primarily motivated by their quest for fame? Are conformist experts obsequiously courting established authorities as a way to promote their own careers and publication records? To unravel some of these complexities we can explore the various reasons that experts might have for promulgating a consensus or a contrarian opinion.

Let's start by looking at some of the incentives driving self-interested experts. In the context of copycat experts, the economic models of self-interested herding that we introduced in chapter 1 assume that individuals are genuinely trying to discover the truth about a situation. This assumption is not unrealistic. Most researchers and scientists are keen to promote the development of knowledge. But what if experts face incentives that create a dissonance between what is best for them as individuals and what is best for society at large? What motivates the selfish expert? Identifying the truth in expert opinions becomes even more complex when we allow that incentives do not always necessarily align with testing the robustness of other scientists' results. When it comes to herding, problems emerge when experts follow a consensus opinion or judgement for reasons that have less to do with the objective pursuit of truth and more to do with their individual motivations, both intrinsic (reflecting personal satisfactions, as we shall see) and extrinsic (primarily the standard economic incentive of money).

Information distortions

Essentially, expertise is about information. A key problem of expertise is not only the absence of clear information, but also problems of distorted information. Information is often not evenly distributed. We don't all know the same things, and often people, including experts, have an incentive to deceive. When we go to experts it is because we are ignorant in some way – we are vulnerable when experts exploit their specialist knowledge. Very many economists have explored the issue of asymmetric information and the problems that emerge when experts exploit their expertise for personal gain. In a more general context, another economics Nobel laureate, George Akerlof, explored some of the consequences of this asymmetry in developing his principle of *adverse selection*, which explains how adverse outcomes and outputs come to dominate a market. Akerlof gave the example of the market in second-hand cars. Because most of us have very limited mechanical knowledge, a used car dealer may exploit our ignorance to sell us a 'lemon' (a dodgy used car). A problem emerges. Not all used car dealers sell lemons. Some sell 'plums' (high-quality used cars). But, because we as the buyer cannot tell the difference between a good car and a bad car, we are only willing to pay for a plum what we would pay for a lemon – a price reflecting the average quality. This is great for those selling lemons, but not so great for those selling plums. From the plum-sellers' perspective, there is not much reason to keep their cars in a market when they cannot get a fair price. They withdraw their plums from the market, the quality of used cars declines and prices fall as average quality falls, meaning more good cars are withdrawn from the market, and the quality and price fall again, and so on. This type of market selects adverse outcomes – that is, the market floods with lemons.[24]

What has this to do with expert opinion? Particularly when the popular press is involved, the quality of expert opinion may be driven down in a similar way. Even if an expert can come up with an accurate judgement, how can we tell the difference between the genuinely knowledgeable expert and the self-promoting expert who is mainly interested in getting their soundbites quoted to advance their own career prospects? It's not easy, and much of the time it's about more than being right or wrong. The truth is that we often cannot tell the difference between a reliable expert who takes great care to research a topic thoroughly and analyse evidence using rigorous methods, and an unreliable expert who might be sloppy in analysing and reporting their data. If the public cannot tell the difference, then each expert may be as likely as the other to get airtime and interviews. So, there are fewer incentives to be reliable, and the quality of expertise declines.

Another type of asymmetric information that experts might opportunistically exploit is *moral hazard*. This problem captures the fact that the incentives of what social scientists call a principal (someone who wants to delegate a task to another) do not always align with the incentives of the agents (someone to whom the task is delegated). This idea is applied across a wide range of economic contexts including labour markets, insurance markets and financial markets. It can be applied to experts too. Whereas adverse selection is about choices we make before signing a contract, moral hazard is a post-contractual problem: when a principal hires an agent to deliver goods or services, they cannot be sure that the agent is not shirking their responsibilities. Agents may have incentives to behave in opportunistic, amoral ways. In the context of experts, we indirectly hire experts and researchers as our agents in the search for knowledge. We, as the experts' principals, cannot easily observe or judge the quality of our experts' output. This creates problems if the experts' incentives do not match our incentives – for example, if they can acquire personal benefits from promulgating eye-catching and newsworthy scientific results. As our hired experts have superior information and it is costly and difficult, if not impossible, for us to monitor their output effectively, then we may be hoodwinked. Expert financial consultants illustrate the problem. Their job is to provide expert financial advice but their personal incentives may instead encourage them to promote particular financial products. Their principals are the recipients of their advice – people who need help with selecting pensions, insurance plans, mortgages or loans – and they will not have the time or expertise to judge the advice they are being given. We may be encouraged to buy a financial product that is not good value or which does not suit us, because we trust an expert even if we cannot judge their expertise.

Moral hazard and adverse selection also apply to experts in other ways, reflecting the fact that experts can conceal the quality of their research findings. Whilst deliberate fraud is rare, there are a few examples of experts who have exploited others' ignorance for their own advantage. One example is Andrew Wakefield, a medical doctor who was first lauded and then vilified for his expert opinions on the combined measles, mumps and rubella (MMR) vaccine. In an article in the esteemed medical journal *The Lancet*, Wakefield claimed that MMR vaccine uptake was implicated in the development of autism and gastrointestinal disease. His opinions hit the headlines and rapidly spread widely, with the consequence that many parents were scared to immunise their children with the MMR vaccine. The problem was not only that these individual children were now susceptible to serious infectious diseases, but also that whole communities became vulnerable to them. Herd immunity – when everyone in a population is protected from infectious disease because a large proportion are immune – was threatened. As with the instability in financial markets that we explored in the previous chapter, the actions of a lone individual spread quickly and widely through complex social systems, generating instability, which is exacerbated by herding. Other researchers tried to replicate Wakefield's findings but they could not. His peers concluded that his paper about the consequences of MMR vaccines had been based on falsified evidence. *The Lancet* retracted his paper, and Wakefield was later struck off the UK's medical register. Why would he have taken this risk with his career? The British journalist Brian Deer investigated the case for an article in *The Sunday Times*, later published in the *British Medical Journal*. Deer claimed that Wakefield had been motivated by his own interests – he had allegedly been hired by lawyers in a lawsuit against the MMR vaccine's manufacturers.[25] If this is true then financial incentives and Wakefield's own self-interest had overwhelmed the moral principles that we expect our medical doctors to uphold, but this was only possible because of asymmetric information.

Reputation

Economists Matthias Effinger and Mattias Polborn explore how herding and anti-herding both reflect an investment in reputation by experts. Experts will realise that they can significantly build their reputations if they are the only 'smart' expert – the only one who gets it right. The herd may be right, or it may be wrong. The point is that, if the herd turns out to be wrong, then being the only smart expert who is right can reap large rewards in terms of money and/or reputation, whereas the benefits of being correct alongside others are less. Anti-herding is therefore more likely when there are large rewards from being the lone smart expert. Then, experts will have an interest in contradicting the expert opinions of other experts. However, reputation can also be susceptible to herding. As we've seen in previous chapters, in many circumstances, our reputations survive better if we agree with the group. We are less likely to be contrarian because we face disproportionate losses if we are dissenters. We take fewer risks with our reputation if we conform, a point introduced in the context of self-interested herding in chapter 1. If an expert has invested years of their career in a specific theory or position, then it is not surprising that they resist change or dissent.[26] As illustrated by the Squier case described earlier in this chapter, there are large costs in terms of career and reputation for experts who disagree with a consensus.

Experts may also tend towards myopic consensus. Agreeing with a crowd may be helpful in building a research career in the short term. However, it is less likely to yield career rewards, in terms of original research and insights, in the long run. But short-term impact may be particularly pressing for young experts at the start of their careers. In building their own reputations, junior members of a research lab often imitate and follow their mentors – partly reflecting social learning, but also because of social pressures. A young researcher who has just received their PhD is more likely to get a tenure track job if they flatter their seniors and group leaders by following in their footsteps. This is not necessarily undesirable: juniors may have much to learn from their seniors. In terms of one's career, though, there is more to be made out of being genuinely original – but the associated risks are high in the short term.

In September 2011, the social psychologist Diederik Stapel was suspended by his employer, Tilburg University, for inventing data on the sociology of urban environments. He manufactured evidence that he claimed demonstrated the link between disordered, littered environments and discriminatory behaviour and deprivation. He sustained his academic fraud for some years because those who suspected he had falsified the data felt unable to challenge him. Stapel reportedly responded aggressively when others, especially junior researchers, questioned his data and findings.[27] This demonstrates the pressures that most of us feel to agree with a group. A junior researcher who disagrees with their seniors, and the whistleblowers who publicly reveal their concerns about falsified or misleading data and analysis, stand to lose all the personal capital they have invested in their careers and networks. They may be ostracised by their bosses and find that their careers stall without the support of a powerful mentor.

However, an expert cannot build a good reputation, at least not in the long run, if they are manufacturing evidence. What motivates people like Diederik Stapel to take such extreme risks with their reputation and their careers? For most experts, there are rewards to contrarianism. As we saw in chapter 5, maverick contrarians are more inclined to take extreme risks than conformist copycats. Added to this, the research community values originality particularly highly. A researcher who just agrees with others may start to incur costs in terms of slowed career progression due to their safe but unoriginal research strategy. Experts ambitious about building their reputations may have an incentive to invent startling findings if these can give them a reputation for original thinking, and their junior colleagues may be scared to dissent. For the copycat experts, their susceptibility to group influence can have profoundly negative consequences, especially if group leaders can exploit their juniors' obedience to authority to manipulate the path of research.

Experts in equilibrium

How can we pull all these elements together into a model that captures the social influences on experts whether they be conformist or contrarian? It can partly be understood as a process of balancing benefits and costs, broadly defined. Most experts, consciously or unconsciously, will focus on the private value of their personal beliefs and opinions. They value truth, but they are also subject to other intrinsic and extrinsic motivations. Researchers may have many friendly chats in the pub with their peers if they are generally in agreement with them. Their senior colleagues may invite them to participate on a research team if they are impressed by their aptitude. There are psychological benefits from conforming to others' beliefs; being contrarian is a far more isolating strategy. Also, experts may gain strategic advantages from joining a group. This links into what economists call *payoff externalities*. When an expert contributes to a growing consensus in a particular direction, this accelerates the movement of others in that direction. The rewards for those who join the herd increase as others join, and then decrease depending on the number of others joining the consensus. When an original, innovative view is taking off and an expert joins a small, elite group who hold it, the value of joining that group increases. Something like a knowledge bubble is generated. As the group grows larger, other rewards kick in. Reputation grows, conformity with others is satisfying. Strategically, joining a new consensus has career benefits. The consensus grows as experts replicate other researchers' novel findings, though to individuals the value of replication can be small – a particular problem for academic research. But it is not a linear process. Once the consensus has taken hold and no longer seems novel and original, the returns for joining the consensus start to decrease. Publishing ideas around an established consensus becomes difficult because the findings are no longer original. As the consensus-forming group swells, then each new expert joining this group gains less and less. In economists' language, the marginal returns from joining the consensus group will fall. Eventually, these marginal returns may reduce to zero, for example if supporting a consensus view is deemed unoriginal and judged to contribute little to the development of new research ideas. There may be stagnation, lots of reinventing the wheel and, at best, insignificant and marginal accretions of knowledge. Then, an ambitious researcher will have nothing to gain from joining the consensus.

The contrarian researcher's rewards come from the opposite direction: as more experts join the consensus, the more of a pariah the contrarian will seem. The contrarian expert will be the loser from the knowledge bubbles that develop as herds of experts follow and develop a new consensus opinion. The contrarian's reputation will falter and their career will stagnate. Eventually, though, the balance may shift. As the consensus view starts to seem unoriginal, the rewards for holding the contrarian view may still fall, but at a decreasing rate. They may even start to rise again as everyone gets fed up with the consensus view, more information comes along and a paradigm shift turns the contrarian into a trendsetter. There is a stable point, an equilibrium, when the gains from consensus and contrarian viewpoints are balanced.

Expert bias

The influences outlined above are largely objective and conscious. More intractable problems emerge, especially in uncertain situations, when experts unconsciously use herding heuristics and other rules of thumb to guide their interpretation of events. Their beliefs coincide with the prior opinions of others, and their private judgements are lost. This is not about individuals pursuing their own self-interest in career or other terms. Instead, unconscious biases are leading experts down the wrong path. Whilst social influences are less benign when experts consciously manipulate them to protect and build their reputations, at least these conscious transgressions can be controlled, for instance via cleverly designed incentive structures, or via sanctions and punishments. If experts' judgements are distorted without them even realising it, then that is a harder problem to solve.

As we have seen in previous chapters, in understanding the role played by psychological factors in our decision-making, behavioural scientists are exploring how and why people use quick decision-making rules – heuristics and rules of thumb – when they are faced with complex information. As Daniel Kahneman and Amos Tversky observed, heuristics and rules of thumb lead to bias, including group biases such as groupthink, which emerge when an individual's beliefs coincide with prior opinions of others around them for reasons that are not objective. This creates herding and *path dependency* – the future is determined by the past, rather than a comprehensive assessment of current, up-to-date information or what is new and different. Sociopsychological influences compound these problems – for example, many of us feel more comfortable conforming. A bias towards herding may also reflect work pressures. For example, one study found that around 78 per cent of Spanish doctors treating patients with multiple sclerosis were likely to follow the herd in recommending treatments. The researchers identified mental fatigue in the context of cognitively demanding decision-making as a key factor.[28] Related to herding bias is the problem of confirmation bias. Behavioural economists and psychologists have shown that people tend to interpret evidence to support their own world view. For example, if a person is a climate-change denier, then they will tend to interpret evidence about the slowdown in global warming as supporting their prior beliefs – that is, as a sign that climate change is a myth. Confirmation bias will affect people's opinions of experts and expert evidence, and so group beliefs and herd opinions will persist.

Researchers have explored the extent to which this sort of phenomenon operates in scientific research too. One example is the Sokal hoax. In 1996, the physics professor Alan Sokal decided to test the refereeing process for academic journals. He submitted a nonsensical research paper – 'Transgressing the Boundaries: Towards a Transformative Hermeneutics of Quantum Gravity' – to the research journal *Social Text*, structuring his fabricated nonsense around prevailing opinions in the social sciences. His contrived paper was accepted by the journal and its referees. According to Sokal, this was because it fitted well with the journal reviewers' and editors' preconceptions. It confirmed their world view and so they were willing to accept it.[29] Experts do of course make genuine mistakes. But they may check for errors more carefully if their initial findings conflict with their prior opinions than if they do not – giving an additional foothold for confirmation bias. Shortcomings in research methodology can be downplayed. When researchers are prone to unconscious bias, they may genuinely believe that their evidence has a strong objective basis when it does not.

Another behavioural bias relevant to herding and social influences reflects 'anchoring and adjustment' heuristics which, as explained in chapter 3, were identified by Kahneman and Tversky.[30] Behavioural economists and economic psychologists have shown that many of our decisions are made around reference points: we anchor and adjust our decisions relative to the status quo. Social influences are important in this because many of our reference points are socially determined – we are naturally biased towards popular existing opinions. Another insight from the literature on heuristics and bias that may have relevance is a problem we explored in earlier chapters, that of loss aversion, as also identified by Kahneman, Tversky and others. The psychic and practical losses to reputation from disagreeing with the consensus are potentially disproportionately large relative to the gains from conforming, and, in a world in which we are more prone to worry about losses than gains, we are more likely to see experts avoiding the reputational risks they would be taking by dissenting.

The personalities of scientists also determine their tendencies to be copycats or contrarians. Strong personalities may be more likely to hold strong convictions – but are such people less likely to herd because of those convictions or because of their strong personalities? How do we unravel the two in our search for truth? In the experimental sciences, we often imagine that careful design of clean experiments and/or the robust application of statistical principles and the scientific method can limit the chances of blind groupthink. To an extent they can, if a researcher has insight and self-awareness. But statistics can be manipulated to persuade, and confirmation bias is hard to overcome, even amongst the most insightful researchers.

Experts' herding externalities

These influences on individual researchers have wider impacts beyond the individual expert. All of us want to do well for ourselves, even if we moderate this with philanthropic inclinations. The problem is that experts' judgements, by their very nature, have implications for other people. These are a type of 'externality' – the term that economists use to describe the costs or benefits an individual imposes on others around them, when these others have no control over the individual's choice or decision. Specifically, groupthink and herding may help a lone expert but generate negative externalities for scientific communities and society at large. As copycat experts follow each other, then they are effectively discarding their private knowledge, and society suffers as a consequence.

In chapter 1, we made the point that the negative consequences from herding are not just about whether the herd is going in the right or wrong direction, but about the fact that private information and judgements are lost. We can illustrate this point more clearly in the context of experts. Herding externalities can be a serious problem for scientific research if it means that experts are less likely to discover something new. In the context of experts' opinions, missing new insights can reflect the excessive weight assigned to a theory that is popular. An individual researcher may find evidence that contradicts the consensus and, for a range of reasons, may discard it. A financial analyst assessing the prospects of an investment in the subprime mortgage market, for example, may have a hunch that these assets are toxic, but they see others around them continuing to invest in them. They weight this evidence more strongly than their own private judgement about the risks of investing in these assets. A speculative housing asset bubble grows, with devastating consequences for people and economies across the world.

Experts in the crowd

Developing this theme leads us into some ideas from economists about the disconnect between what is best for the individual and what is best for the group. Knowledge and evidence can and should share many of the characteristics of what economists call public goods. In their purest form, public goods are fully accessible to everyone. Individuals are not excluded from consuming them. There are no barriers to entry. One person's consumption of them does not diminish the potential for others to consume them. The stock of public goods in their purest form does not deplete, and the marginal cost of one more person using them is zero. From an individualistic perspective, the problem of public goods is that there is no market incentive to provide them, given that it is difficult to charge people if you cannot easily stop them from consuming. And if you cannot charge people, you cannot make a profit. So, who pays for public goods?

From a societal perspective, amassing knowledge is a collective effort, and institutions other than markets have evolved to support this, though market institutions have also evolved to make a profit from it. Most controversially, in academia, profits are made by effectively privatising knowledge via scientific journals' expensive paywalls and/or financing arrangements in which the academic researchers themselves are charged for publishing their own research. Specifically in the context of copycat experts, the collective nature of research and knowledge accumulation makes it hard to separate consensual beliefs that are well grounded from consensual beliefs that lack proper foundation. If accumulating knowledge is a collective effort by large numbers of experts, then no single expert can be held responsible for errors.

As we noted above, reputation is affected as the balance favouring copycat experts shifts in favour of contrarian experts. When copycats' reputations are more robust, consensual beliefs will generate over-consensus and group bias. Empirical philosopher Michael Weisberg and his colleagues have explored the idea that consensual beliefs have negative impacts at an aggregate scale. Using computational modelling methods, Weisberg and his team artificially generated two types of population, one dominated by copycat 'followers' and the other by contrarian 'mavericks'. They created visual maps to capture how much of a knowledge landscape was explored by either group. Their simulations showed that substantially more ground was explored by mavericks than by followers. Followers explore less because they are sticking with the crowd. Mavericks explore more because they venture into territories where others haven't yet been. The implication for experts is that if an expert community is dominated by large groups of followers, then the knowledge landscape is not fully explored. Experts who are followers learn much less when they are all copying each other. In epistemic terms, essentially, they are just retreading ground already well trodden by others. With a good proportion of mavericks in a population of experts, the outcome is reversed. The knowledge landscape is more likely to be fully explored. Experts are more likely to discover more when they focus less on what their predecessors have explored. So, Weisberg advocates incentives for risk-taking in research – to overcome the welfare loss from too many copycats just imitating each other.[31] Weisberg's study shows that contrarians are essential. We need contrarians to shepherd herds of experts away from a path dominated by social influences, towards fresh perspectives and new interpretations of data and evidence. There are no easy answers, though, because, social influences can be valuable too – for example, replicating results is an essential but neglected aspect of scientific research. If a hypothesis has genuinely been verified across a range of different studies then that may be because it is a more plausible and probable hypothesis than the alternatives.

As we have seen, economists' models of herding show why we might logically follow others if we believe they have better information than we do. By extension, supporting consensus views does not necessarily mean that those views are wrong. It may be logical to ignore what little we know already if we can do better for ourselves by following others. This is true for experts too. The problem is that, at a macro level, it leads to path dependency. This insight can be simplified to the observation that if more experts support a theory then, all things being equal, perhaps it is more likely to be true. That does not mean that it is a definite truth. Academic research is not generally about absolute proof. Imagine two competing hypotheses, both of which are initially novel and have no 'tribal' support. When a theory or hypothesis is widely supported by many experts then it is reasonable to believe that it is more *likely* to be true. The chances of a large number of experts supporting a false hypothesis may seem smaller than the chances of a large number supporting a true hypothesis, especially when experts have good, objective reasons for agreeing with each other.

Consensus rarely holds for ever. As we saw above in the examples of shaken baby syndrome and stomach ulcers, contrarians come along and shift opinions. As the philosopher Thomas Kuhn asserted, the evolution of knowledge is generally peaceful. But when learning and knowledge are partly formed by social interactions, not just supported by them, then knowledge can go down a very crooked path. If we were always conscious of this, it might not be such a problem; but academics and other experts may, consciously or unconsciously, follow a group consensus. It is lucky, then, that, as Kuhn observed, there are intermittent revolutionary phases accompanying paradigm shifts – when the consensus is suddenly jolted onto a new path.[32]

Amateur experts

Some of the lessons from experts can be applied to collective decision-making more widely. Experts are not the only ones susceptible to these individual pressures when forming opinions with consequences for others – juries are an example. The balance of positive and negative externalities will intensify as opinions spread from experts to amateurs, because amateur opinions have weaker foundations. By definition, amateurs do not have deep knowledge and expertise. They have less private information to use, and this makes them more strongly susceptible to social influences. Sometimes herding is the only obvious option when information is very patchy and uncertainty is endemic. On juries, judgements concerning guilt or innocence are often influenced by group dynamics and herd behaviour. Mock jury experiments illustrate that social influences can generate significant distortions.[33] Mock jurors are susceptible to peer pressure and some studies have shown that the susceptibility of an individual juror to the opinion of others is affected by individual differences in personality. The fact that some types of juror are more conforming than others suggests that some jurors may be more easily influenced than others. Juries are not as impartial, objective and uniform as we need them to be.[34]

Another example of amateurs thrown into a quasi-expert role are the lay members of investment oversight committees. Dr Anna Tilba from Durham University and I were commissioned by the UK's Financial Conduct Authority to explore some of these group influences. We focused on the impacts they might have in impeding competition in the UK's asset management industry. This industry includes the large institutional investors – pension funds, insurance companies, charities and endowment trusts. Together they manage large portfolios of funds, and oversight committees are constituted to ensure that this job is done well. We focused particularly on pension fund oversight committees, which manage very large sums of money. Amateurs, such as employee representatives, are included on the investors' decision-making committees. A key task for these committees is to appoint investment consultants, and they often do this by way of a 'beauty parade', whereby different investment consultants present what they could offer the investor, and the committee decides who to choose.[35] Herding can have a strong influence during these parades. Imagine that one member of an oversight committee has superior private information. They may have done extra research into the options available and/or the track records of the investment consultants interviewed by the committee. But when this committee member sees how other members are deciding, especially more senior members and/or other members to whom they attribute superior decision-making capacities, then they will often defer to the group decision. Amateurs are often included on these committees because a diversity of opinions is valuable. If, however, these amateurs are too easily persuaded to go with the consensus view, then the value of their representation will be lost.[36]

All experts – whether scientists, academic researchers or expert witnesses such as doctors and lawyers – are generally keen to give their objective view, based on truth. We need people to interpret evidence for us. In reality, poor information, unreliable data and profound uncertainty mean that it is not so easy to untangle the evidence. We cannot always separate good hypotheses and theories from bad. When the lacunae in knowledge or understanding are large, experts become as susceptible to herding influences as anyone else. As we have seen, this social susceptibility can have profound implications, not only for the individual expert, but also for the path of knowledge and research. The wider social costs can be large.

Whether experts are copycats or contrarians, interpreting their opinions can be problematic. We cannot know whether a copycat or a contrarian opinion is better. Expert opinions may be distorted by consensual experts herding behind the consensus view because they find it easier, or by contrarian experts promulgating a divergent view for the sake of attention and career progression. We need contrarian experts, but we need them to be contrarian for good reasons. The challenge is to separate the motivations and incentives that might lead an expert mistakenly to agree with a herd consensus from a genuinely supported consensual view that is correct because the consensus is correct. Similarly, in interpreting a contrarian expert's views, the challenge is to balance the extent to which contrarians' personal motivations and incentives are driving their opinion against the extent to which their contrarian view is more likely to be correct because it is challenging a misplaced consensus. The herd consensus may be right, it may be wrong. In interpreting expert opinions, our Herculean challenge comes in telling the difference. When experts find it difficult to interpret evidence, then they will be less sure that they have the correct answer. And, as we have seen in previous chapters, herding is more likely to take hold when people are unsure. An expert with strong convictions may be less susceptible to blind conformity, but if their strong convictions reflect overconfidence then their dissent may be as destructive as being excessively susceptible to collective opinion.

So, what can we do about it? Some solutions may lie in developing some parallels with the literature on the management of common resources. Nobel Prize-winning economist Elinor Ostrom explained how close-knit communities manage common resources well, much better than many economists predict.[37] Are there lessons for research and knowledge management, tailored to getting the best from experts whether they be copycats or contrarians? We cannot rely on individuals to manage knowledge and expertise because their incentives and biases can lead them, consciously or unconsciously, down a wrong-headed path. We need institutions to ensure the safe stewardship of expertise.

What concrete solutions could be introduced? Communities of researchers and experts should encourage the extensive replication of results. Academic communities could move away from the idea that only novelty and originality have value and are worthy of publication. Professional societies are already developing initiatives in this spirit, such as the Association for Psychological Science's 'Registered Replication Reports' policy. Similarly, journals such as the *Journal of Negative Results* can play an important role in controlling the fads and fashions in academic research driven by an ephemeral preoccupation with attracting public attention. Journal editorial boards can limit the influence of social pressures by ensuring the anonymity of journal submissions and blind reviewing of submissions. We need more academic, scientific and professional institutions that encourage dissent. Even-handed monitoring of researchers, publishing the names of a paper's reviewers alongside the paper, and requiring researchers to publish the data they have used to justify their conclusion can all help. Some journals and learned societies have instituted these solutions already. But it will be difficult to implement practical concrete initiatives if professional associations, expert groups, journals and publishers have too much invested in the status quo.

In this chapter, we have seen that social influences have more traction in uncertain situations. How do we judge our experts? They can be all combinations of good and bad, and right and wrong. Often, we can't know which. We know that Marshall and Galileo were correct. Stapel and Wakefield were wrong. The problem remains that we cannot judge very easily if we do not know the truth. But, as noted above, we can implement practices to ensure that we get as close to the truth as possible as soon as possible. Better standards of analysis and reporting in academic research, better and more transparent journal review protocols, better education so that lay people can more easily understand scientific arguments, clearer sanctions on experts who exploit their authority over others: all these solutions could help to ensure that distortions reflecting the unconscious biases of experts, whether copycats or contrarians, are minimised.

A feature we have seen in this exploration of experts and their opinions is that a forceful personality can often distort experts' assessment of evidence. These distortions can be especially large if a vigorous, aggressive personality leads a group. Group members will, understandably, be reluctant to dissent from the views of the leader, either for psychological or economic reasons. The relationship between leaders and their followers illustrates again that copycats and contrarians do not exist in isolation, but are enmeshed in a symbiotic relationship. We shall turn to this relationship between contrarian leaders and copycat followers in the next chapter.

8

Following the leader

In the well-known fairy tale 'The Pied Piper of Hamelin', a rat-catcher is hired by the mayor of Hamelin to deal with the town's rat infestation. Playing his magic pipe, the Pied Piper entices the rats away from the town and drowns them in the river. When the mayor refuses to pay the rat-catcher he punishes the town by luring its children away, following him and his music into the mountains. It is a strange and wonderful story, though there may also be some truth in the tale, with some accounts suggesting that it concerns the deaths of children during the plague.[1] Whatever the case, it is an intriguing example of a leader's power over their followers.

Another all-too-real example is provided by today's global terrorism. On 11 September 2001, nineteen al-Qaeda terrorists led four coordinated aircraft attacks on New York, Washington and the Pentagon. The attacks caused the deaths of close on 3,000 civilians, with many others injured, as well as trillions of dollars of damage to property and infrastructure. This event is burnt more indelibly on our collective memory than any other in recent times. The motivations of al-Qaeda's founder Osama bin Laden and his confederates seem to be straightforwardly apparent: they gained power and some gory glory from the event. The question that seems unanswerable to many of us is: What led those who directly perpetrated the attacks to obey their leaders in sacrificing their lives in such a spectacular way? This is not a phenomenon limited to the religiously fervent in today's War on Terror. Pressure to participate in horrific acts – from discrimination to genocide and everything in between – is dispiritingly regular in human history. Not even major atrocities such as the Nazi Holocaust and Stalin's purges are as uncommon as we might hope. Genocides are an enduring feature of our history, including those that happened in Rwanda, Bosnia and Darfur not so long ago, and in Iraq and Syria today.[2]

One of the most potentially sinister facets of herding is the relationship between a particular type of contrarian – a leader – and a particular type of copycat – a follower. The interactions between these leaders and followers can have large impacts, positive and negative. A leader's influence can be detrimental on a catastrophic scale. Many brutal dictators have committed horrific crimes against humanity, demonstrating the terrible consequences that can emerge when people blindly obey a despot. And, on a lesser scale, we are surrounded by ambitious politicians manipulating voters in their personal pursuit of power. More edifyingly, some of us also have opportunities to follow benign, egalitarian and benevolent leaders such as Mahatma Gandhi, Martin Luther King and Nelson Mandela, to name a few of the most famous. When we are led by inspiring leaders, the consequences can be as positive as the consequences of following brutal leaders are bleak.

What motivates us to follow a leader? Why do some people demonstrate extreme manifestations of loyalty? For

Sigmund Freud, whose analysis of group psychology we explored in chapter 2, these leaders are essential to our group relationships, especially in institutions such as the army and the Church.[3] For Freud, leaders play a transformative role:

> All the members [of a group] must be equal to one another, but they all want to be ruled by one person. Many equals, who can identify themselves with one another, and a single person superior to them all – that is the situation that we find realised in groups which are capable of subsisting. Let us venture, then, to correct [the assertion] that man is a herd animal and assert that he is rather a horde animal, an individual creature in a horde led by a chief.[4]

Business leaders and followers

The maverick entrepreneurs we explored in chapter 6 provide a simple example of leaders in the economy. They often lead the way in producing and distributing innovative products and services, with other businesses following along behind. Imitation is a common strategy in business, and it can be a good way to maximise profits. Joseph Schumpeter, whose ideas about innovation and entrepreneurship we also introduced in chapter 6, explored how businesses' decisions to imitate each other play out in leader–follower relationships. For Schumpeter, these entrepreneurial leaders are essential to a thriving economy. Innovative, risk-seeking entrepreneurs lead swarms of imitators and so play an essential role in catalysing new waves of business activity.[5] At a microeconomic level, leader–follower relationships are easy to explain in terms of self-interest and can be understood from the relatively simple perspective of rational choice theory. German economist Heinrich Freiherr von Stackelberg captured leader–follower relationships in his model of industry leadership of oligopolistic firms (a classic example in undergraduate economics textbooks).[6] Stackelberg's model is used to illustrate what economists call a *first-mover advantage*. If a new business produces something innovative, or perhaps just moves into a new area that is currently lacking a product or service, it takes the advantage of being first on the scene and mops up most of the potential customers. Latecomers are left with just the small number of customers remaining.

To illustrate, imagine a small town that is not yet connected to broadband. An internet provider spots the opportunity and wants to enter the market. To do so the provider needs to invest a lot in terms of start-up costs, new technology and new infrastructure – these are examples of what economists call *barriers to entry*. When barriers to entry are high and costly, it is hard for new businesses to enter a market because they have to spend so much to get started. For the first provider to enter the market, the revenue and profits may justify the costs of entry into the market. But if a second internet provider considers entering the market, they too would have to overcome the same barriers and invest in new technology and infrastructure – but for much less revenue if most potential customers have already signed up to the first provider. The business case for this second mover may not be strong, so they may decide not to bother. These first-mover advantages are one reason why monopolies and oligopolies face so little competitive pressure to bring their prices down to a level consistent with consumer welfare, and this is why these types of industries are often regulated.

But leading businesses do not always enjoy a first-mover advantage. In other situations, perhaps where the business model, product or service is more complex, a follower can learn from the leader and improve their business strategies accordingly. Then the follower will be able to enjoy a *second-mover advantage*. Here, the follower wins. Drug design by pharmaceutical companies is a contentious example. One business invests money in research and development to develop a new pharmaceutical. A follower can come along and free-ride on the investment and technological innovations of the first business, offering a generic medicine at a much lower price and thus capturing a good chunk of the market. Partly, this is a good thing for consumers, particularly in the developing world, where people are urgently in need of access to cheap pharmaceuticals. If followers can take away a good chunk of your profits, though, what incentive is there to be a leading innovator? The first movers therefore protect their innovations with patents. The general point is that either leaders or followers – first movers or second movers – can be winners in a simple economic world. Some successful entrepreneurs and speculators will be aware of when it works to be a leader and when it works to be a follower. They will build these insights into their business strategies, swapping roles when there is a suitable opportunity.

Economic theorist Harold Hotelling presented another microeconomic perspective on copycats in the business world in his simple model to explain why businesses copy each other in deciding where to locate their premises. Imagine that there are two ice-cream sellers, Ben and Jerry, on Bondi Beach. You would think that each would locate themselves a long way away from each other so as not to be competing for customers. Hotelling's model shows that the ice-cream sellers will in fact capture a much smaller chunk of the market if they are far away than if they are close together, and so both sellers will move until they are as close to each other as is possible. Let's say that Ben has already set up his ice-cream stall in the middle of the kilometre-long Bondi Beach, and Jerry sees he's doing a roaring trade. We'll also assume that the potential customers are all lazy beach bums and will just go to their closest stall to buy their ice creams. Where should Jerry set up his business? If Jerry decides to set up shop 200 metres south of Ben, he will attract a total of 400 metres-worth of customers: 100 metres-worth to the north (i.e. half the customers between Ben and him, because the other half will still be closer to Ben's stall and buy their ice creams there) plus all the customers south of his own stall – another 300 metres. Ben will do much better: he will get all the customers north of his stall (500 metres-worth) as well as the 100 metres-worth of customers between himself and Jerry – a total of 600 metres-worth of customers. But then Jerry thinks: what if I set up right next door to Ben? Jerry will then capture all the 500 metres-worth of customers to the south of the two stalls, and Ben will keep just 500 metres-worth of customers to the north. Jerry will maximise his profits, and take half of Ben's, by locating himself as close as possible to Ben. Hotelling's model helps to explain why we often see similar shops – takeaways, betting shops, clothes retailers, estate agents – all collected together in one area of our high streets. Businesses copy each other with their business location decisions and, in imitating other businesses, business leaders capture markets and customers.[7] Political scientists have also borrowed this insight to formulate the *median voter theorem*, explaining why political parties will try to pitch their manifestos to the median average consensus view to gain the majority vote share – though this insight seems less enduring in today's more polarised political landscape.

In some places, for example cities, formal and informal forms of 'place-based' leadership are a key determinant of economic growth. Regional scientists Andrew Beer and Terry Clower have unravelled some of the roots of what is called 'place-based' leadership, that is, leaders who represent specific places – for example, local communities, local authorities, cities, regions and states. Effective leaders can help communities and regions to form and implement a vision of what they want for the future, monitor the progress of policies, and adjust strategies when those policies are not turning out as expected. Place-based leadership can take many forms depending on the organisational context. Sometimes leadership is formalised within traditional hierarchies and formal roles, such as city mayors. Other times, leadership can be more informal, such as community-based leaders. Informal leadership often leverages 'slack resources' – people who have the time and energy to volunteer, for example in representing their communities on advisory committees for regional development agencies. No one type of leadership is more important than the other. Both formal and informal leadership are essential to a region's success. And regions need a diversity of leaders. As for any leadership, the personalities of place-based leaders do not necessarily fit the stereotype of a loud and gregarious 'great leader' who leads by talking. Undoubtedly, 'loud leaders' can be better at building networks and contacts, but 'quiet leaders' also have distinct qualities: they lead by doing, and focus on building trust and effective collaborative relationships. All these qualities contribute as much to leadership as extraversion and an imposing personality. Overall, Beer and Clower conclude that effective leadership is indispensable at a local and regional level. Places with good leaders are more likely to succeed economically because economic performance is no longer so dependent on whatever resources may be available in a local area. Building infrastructure and attracting entrepreneurs and skilled workforces are all irreplaceable. Budgetary constraints are crucial too: in countries where government expenditure is centralised, effective leaders can make a difference to how much a specific region is supported by central authorities.[8]

Following neighbours

Consumers' inclinations to follow the leader can be harnessed as an economic policy tool, encouraging us to herd behind others, sometimes helping to reduce the external costs incurred when individuals act in their own self-interest while disregarding the economic consequences for the wider economy. Some examples relate to energy and the environment. Leaders can act as champions for constructive social behaviours, facilitating social learning about best environmental practice. In the UK, a group of environmental scientists conducted some experiments to explore workers' environmental behaviours via an 'Environmental Champions' programme. Over three months, Environmental Champions were assigned to lead 280 office-based workers in campaigning, improving environmental information and providing practical advice about how to reduce environmental footprints. The programme was very effective: it led to a 12 per cent reduction in energy consumption and a 38 per cent reduction in waste production in the participants' workplaces. Environmental Champion leaders played a constructive role, inculcating good environmental practices in their followers.[9]

Relationships between leaders and followers play a crucial role in determining our consumption choices too. Most of us get information and ideas through social media via personal contacts – friends, friends of friends and friends of friends of friends, and so on. Focusing on the idea that information is most effectively disseminated via personal contacts, economists Andrea Galeotti and Sanjeev Goyal used mathematical models of social networks to capture the leading role played by 'influencers' – the small group of people who have a large impact on the choices and decisions of others around them, for example on consumers looking for information to guide their purchases. Galeotti and Goyal call this phenomenon the 'Law of the Few': influencers are often leaders prepared to make up their own minds, without needing the reassurance of the herd.[10] Why do influencers have such power over the rest of us? There is no difference between them and us, apart from their dense and extensive social networks. They are connected with many more people than the rest of us. Therefore, information about influencers' choices spreads rapidly around social networks because influencers have so many connections.[11]

In our social media-saturated world, influencers have found their way from economic theory into the real business of fashion. Modern marketers understand well the importance of role models and trendsetters, and the impact they can have via social media. By leveraging our instinct to follow leaders in our consumption choices, businesses can generate a lot of additional exposure and sales by identifying and incentivising fashion leaders with hordes of followers to endorse their brands. So, high-end and high-street retailers are now routinely enlisting influencers from around the world to connect with millions of their followers through their social media networks. For example, when launching their Spring–Summer 2016 collection, the fashion chain Mango's #MangoGirls campaign recruited a selection of female fashion bloggers – specifically those with large numbers of Twitter and Instagram followers – to form a season-on-season relationship with the brand. Similarly, the luxury shoe brand Jimmy Choo has its own group of influencers who disseminate fashion advice and opinions online – always complimentary to the company. Like other fashion companies, it rewards its influencers with freebies and experiences such as #Chootravels – trips to enviably glamorous destinations like Marrakesh, Zermatt and Rajasthan, treating the influencers to keep them onside while simultaneously providing a steady stream of informal advertising. Bloggers and vloggers invited to India by Jimmy Choo had a combined 'follow-ship' of 6.5 million people.[12]

Why follow the leader? Obedience to authority

Economic models capture only a small snapshot of our experiences with contrarian leaders and their copycat followers. In an economy filled with self-interested and rational individuals, leader–follower relationships unfold in relatively harmless ways, as we have seen. But our world is not as simple as that portrayed in the economists' models. Once we introduce sociopsychological influences into the mix, the consequences are not necessarily nearly so benign. A leader cannot lead without an obedient crowd. Followers must be inclined by some power or authority to follow – and social pressure plays a key role not only in sustaining cooperation and mimicry within groups but also in inculcating a follow-the-leader mentality. These social pressures are powerful. We are conditioned to conform not only to a *group* as a whole, but also to the judgements and opinions of *individuals*, including parents and seniors, and later in life our bosses and other authority figures. We conform because the real or imagined pushback we get from peer pressure makes us uncomfortable, as we saw in chapter 2 with Solomon Asch's line experiments. We also conform to the orders or expectations of authority figures partly because of social pressure but also because we fear some sort of retribution.

These social pressures drive obedience to authority, an essential feature of many leader–follower relationships, and plentiful evidence from social science has shown that, just as we have an instinct to conform, so we have an instinct to obey. With his research group, social psychologist Stanley Milgram developed a series of early (and controversial) experiments to test the limits of our willingness to obey authority figures. Milgram and his team wanted to understand why so many ordinary people are often complicit with their tyrannical governments. Milgram was particularly keen to explore the role played by ordinary people in the atrocities committed by Hitler's Nazi government – not only why these otherwise ordinary people were prepared to be abnormally vicious, but also why they seemed unprepared to take any personal responsibility for their actions.

To unravel some of the influences, Milgram and his team set up an experiment requiring their participant volunteers to inflict brutal punishments. How ruthless were ordinary people prepared to be in the process of 'just following orders'? Milgram's experimental participants thought that they had been recruited into a conditioning experiment designed to test how punishment affected learning. The experimenters instructed them to train 'learners' by administering electric shocks each time the learners made a mistake. The participants were told that the intensity of the shocks would increase, from 15 to 450 volts, according to the number of mistakes the learners made. Unbeknownst to the participants in these experiments, they were not administering real electric shocks at all. The learners were really actors pretending to make mistakes and suffer pain. Around 65 per cent of Milgram's participants were prepared to administer what they thought were near-deadly electric shocks of 450 volts when instructed to do so by an authority figure. All participants were prepared to inflict 300-volt shocks. In a variant of the experiment, the participants were given the opportunity to observe other participants – referred to as 'teachers' – who refused to administer the shocks. When the participants had a chance to observe the teachers, they were less likely to obey the instructions to inflict shocks. Overall, Milgram's experiments suggest that leaders have a strong influence on their followers, but peer pressure from others at the same level in a hierarchy can also play a role in modifying blind obedience to authority.[13]

More generally, Milgram's evidence suggests that many situations can be manipulated, by leaders or just by circumstance, so that ordinary people are led to commit egregious acts. This helps to explain why malevolent dictators and others can have so much influence over otherwise empathetic individuals. Nonetheless, we are prone to conflicts between our conscience and our instincts to obey authority.[14] In later work, Milgram set out the idea that obedience reflects a tension between our autonomous states of being and our 'agentic' states of being. In the former, we take responsibility for our own actions. In the latter, we allow others to tell us what to do and we blame them. We lose our sense of autonomy when we become someone else's agent and instrument, but for a leader to dominate our actions we must perceive them as a legitimate and qualified authority figure. We encourage ourselves to believe that our leaders will accept the responsibility that we have abrogated, and sometimes we blindly assume, without much foundation, that they are leading us in a just and responsible way.[15]

Thinking styles in leaders and followers

Through this book we have explored how interplays between System 1 instinct and emotion and System 2 reason and cognition drive our copycat and contrarian choices and decisions. Does this sort of dual-system processing still hold under the kind of extreme conditions and duress that Milgram's participants experienced? Ethical constraints mean that researchers cannot easily explore these questions with real humans, so a multidisciplinary team of neuroscientists, psychologists and computer scientists from the UK, Austria and Spain found a novel way to circumvent them, using virtual reality technology and brain imaging techniques to explore the neural responses of people involved in Milgram-style experiments. Sixteen healthy adults were recruited and immersed in a virtual reality world. Just as in Milgram's earlier experiment, the participants were instructed to administer shocks – but this time to a female avatar programmed to respond to the 'shocks' by mimicking human expressions of pain. If the avatar gave a correct answer to a question, the participants were instructed to press one button to indicate that they did not want to give the avatar a shock. But if the avatar gave an incorrect answer, then the participants were instructed to press another button to inflict an electric shock, no matter how painful the shock seemed to be.

The experimenters scanned the participants' brains using fMRI. Observing pain in the avatar did activate the participants' amygdala, which, as we have seen in previous chapters, is commonly associated with processing aversive emotions including fear and anxiety. The activations here were consistent with the operation of some fast-thinking emotions. Were the participants sharing the avatar's apparent fear? The perceptions of the avatar's pain also induced responses in the participants' prefrontal cortex, which is generally associated with higher level, slow-thinking responses. So, fast and slow thinking were working simultaneously in this virtual reality version of the Milgram experiments. Whilst the experimental team were not able specifically to assess what was driving the participants' obedience, they were still able to ascertain that participants will persevere with instructions even when they are experiencing emotional distress themselves.[16] If this evidence can be generalised to real-world experiences, then it might suggest that people's decisions to obey the authority of their leaders, for example in following a leader's instructions to inflict pain on others, is not an easy choice, and is associated with emotional conflicts within the psyche of the leaders' followers.

Students in prison

Stanley Milgram and his team's research is a classic of social psychology, and it inspired other social scientists to carry out a range of similar experiments with the aim of further unravelling the hierarchical relationships associated with obedience to authority. A now notorious experiment was the Stanford Prison Experiment, in which students were recruited to participate in an experiment set in a mock prison. The students were given a choice of pretending to be a guard or a prisoner. The experiment soon started to mimic reality closely. The students fell into their roles easily, with the 'guards' exhibiting genuinely domineering and aggressive behaviours towards the 'prisoners'. In turn, the 'prisoners' adopted subservient and submissive behaviours. Everyone, prisoners and guards alike, was complicit in the destructive, antisocial behaviours exhibited by the guards, and the mock prison quickly transformed into a violent and dangerous place – even though all the student participants knew that they were just part of an experiment. Even more worryingly, the experimenters also started to lose their objectivity, rationalising the abusive behaviour of the student guards. The experiment had to be abandoned early for ethical reasons.[17]

The Stanford Prison Experiment demonstrates how strong and ingrained are our tendencies to conform and immerse ourselves in the roles to which we are assigned. But context is not the only driver of dehumanising behaviours. Personality traits also play a role. As we have discussed in earlier chapters, our personalities predispose us to feeling specific emotions – for example, an anxious personality will be predisposed to feel fear. Our individual predispositions will also affect our social emotions – how we feel in social situations when we or others are being treated unfairly, for example. Social emotions will have both positive and negative dimensions, and they may also determine our inclinations to engage in antisocial behaviour. Anticipating that an analysis of personality traits would be illuminating, the prison experiment researchers asked the students to complete some personality tests before the experiment started. They found that students who had chosen to be guards were less sociable, altruistic and empathetic, and scored more highly on tests designed to capture antisocial tendencies, including Machiavellianism, aggression, authoritarianism, narcissism and social dominance.[18] So, the fact that some of the students in the Stanford Prison Experiment were willing to fall into their new roles so quickly might be partly explained by students self-selecting themselves into particular roles, as determined by their predispositions and personality traits.

Suppressing social emotions

Emotions are important in a social context, but they can also be suppressed. Social norms may prohibit the expression of emotions, from relatively mild cultural conventions (fewer eyebrows are raised at noisy, public expressions of emotions in some cultures than others) to the extreme regulations in institutional settings such as prisons. In the latter dehumanised environments, we are more likely to follow others and obey authority figures.

One interpretation of the Stanford Prison Experiment findings is that the students' emotional responses were suppressed.[19] Some evidence of emotional suppression comes from real-world applications of insights from the electric shock and prison experiments. Outside experimental labs, there are many examples of tyranny feeding on our instincts to obey authority, especially when we ourselves face severe hardships and threats. These instincts can help to explain the perverse relationships between leaders and followers that have characterised some of the most barbaric episodes in human history, and why social emotions are suppressed in extreme environments associated with war and oppression. Stanford psychology professor Philip Zimbardo has written extensively on obedience to authority, power relationships and their impact across a range of contexts. Zimbardo was a co-investigator on the electric shock and prison experiments described above. He was also an expert witness for the defence of American military and intelligence personnel on trial for abuses at the Abu Ghraib military prison in Iraq, after photographic evidence emerged in 2003 of the torture of Iraqi prisoners at the prison during the Iraq War. Abu Ghraib was a real-life corollary of the Stanford Prison Experiment, with staff operating under conditions that were gruelling and degrading. Military personnel were living under the real threat of physical retribution for disobeying authority and/or violating group norms. Zimbardo attributed the abusive behaviour of the Abu Ghraib defendants to a 'Lucifer effect', claiming that any of us might have the capacity to be vicious if we found ourselves in such aggressive, dehumanised environments. Most of us have the capacity to act in a way that could be judged evil by others if we are put under enough pressure. The responsibility for acting in this way is not ours alone. Our propensity to be villains (or heroes) is formed by the authority figures and contexts in which we find ourselves. In these situations, we will be driven by our instincts to obey leaders' orders to commit ruthless acts that we would not for a moment contemplate if given the choice under different conditions. Without institutional prohibitions, and in a less degrading context, those caught up in the Abu

Ghraib scandals might have behaved in a less vicious way.[20] And those caught up in these dehumanised situations find ways to control their normal empathetic responses.

In Abu Ghraib, and also more recently in the American prison of Guantanamo Bay in Cuba, military and penal operatives have suppressed their own more humane and empathetic social and emotional responses in committing violent acts. But this is not all about impulses and instincts to obey authority. In the face of potentially violent retribution from their leaders for disobeying or from their peers for rebelling, more deliberative thinking styles associated with self-interest and self-preservation will come into play. The military personnel may judge that they have little choice but to obey given that the consequences for themselves might be so severe. These responses are not irrational. Real-life examples show what the consequences are for those who do not fit willingly into a follower role. In the Abu Ghraib case, some evidence later emerged that authorisation for the abuse came from high up the chain of command and was state-sanctioned. The whistleblower, Joe Darby, was initially reassured that his identity would remain a secret – a promise that was allegedly broken by Donald Rumsfeld, the US secretary of defense at the time. Subsequently, Darby and his family had to be taken into military protection because of threats from others, including from neighbours who castigated him for betraying his fellow soldiers. Going against the actions of the herd, even for the most honourable of reasons, risks ostracism not only by authority figures but also by peers. Like David Kelly, the whistleblower from chapter 5, Joe Darby suffered severe consequences for refusing to comply with the role of obedient follower.

Obedience to authority operates in more benign contexts than prisons and wartime. Hierarchical relationships also characterise academic and scientific research groups, as explored in the previous chapter. Junior researchers are conditioned to follow in the footsteps of their supervisors and mentors, and to respect the authority of these individuals. When times are uncertain, and individuals lack confidence in their own opinions, there is comfort in conforming to the views of an academic 'tribe'. Something like a safety-in-numbers effect is operating. There may also be an element of fear. Disagreeing with seniors may have negative consequences. As with the Emperor's New Clothes, naked or not, from the perspective of pure self-interest it would be foolish to argue with the person in authority. So, obeying authority is not purely an unconscious response. Self-interested logic and deliberation come together with impulses and instincts to encourage followers to obey their leaders.

Leader–follower symbiosis

Copycats follow copycats, but who leads the copycat herds? Often, it is contrarians. Imitation, after all, requires at least two players – the imitators and the imitated, the followers and the leader – and they come together in a symbiotic relationship. Copycats need a leader, but a leader is nothing without their followers, which means that they must give those following them something in return for their loyalty – a sense of either belonging, identity or purpose. We have to be selective with our leaders because having too many would create confusion. But why do so few of us decide to lead and most of us prefer to follow? Leaders are characterised by less of a tendency to herd than the rest of us, for a range of reasons – economic, psychological and emotional. The contrarian behaviours associated with leadership are rarer in our world because we have evolved as social animals. So, copycat followers and contrarian leaders each have distinctive personality traits, and each is driven in different ways by the balance of fast System 1 emotion and slow System 2 reason. These interplays of fast and slow thinking also help to define the nature of the symbiosis between leaders and followers. Our leader–follower relationships can occasionally have devastating consequences, as we saw above, but mostly they come in much more benign forms. We can characterise these different relationships and the extent to which our decisions to join in are deliberative or instinctive and emotional by looking at the spectrum of groups we join – from clubs and congregations through to cults, as we shall see below.

Clubs

Clubs are groups of people with a common interest – in sports, books or losing weight, for example. Club members join together to share in an activity or enjoy activities together. From the perspective of mainstream economics, clubs are generally easy to explain as an example of rational self-interest. Each club member is helping themselves by collaborating with others. Clubs are a form of coalition and joining them and following the club's leader is often a sensible thing to do. The leader helps the club members to achieve their goals together and more easily. Similarly, team leaders play an essential role. What is the incentive for individual members to exert effort if the outputs from joint efforts are to be shared equally? There is a free-rider problem – each self-interested individual will prefer to have an easy life and let the others do the work. Clubs and teams therefore need a leader to take responsibility for coordinating and incentivising the group, and discouraging shirking. Who should lead? One solution is to create the role of what economists call a *residual claimant*. The residual claimant is penalised (or rewarded) in some monetary or other form if the group output is less (or more) than satisfactory. This residual claimant takes the leadership role because they are offered additional private incentives to motivate other team members. Whether in workplaces or among student groups, successful teams are characterised by good leadership.[21]

Slimming clubs are one example from our domestic lives. When groups of overweight people gather together for diet tips, motivational talks and weekly weigh-ins, the club leader plays an essential role in coordinating the club's activities and providing additional inspiration, often very successfully. The effectiveness of having this sort of mutually shared, relatively objective goal means that the relationship between leader and followers can be of the most productive and mutually beneficial type possible. Sports clubs and teams are similarly about the mutual pursuit of a goal that has the capacity to bring satisfaction to the group. Successful slimming clubs also illustrate some of the interactions between System 1 and System 2 thinking. Psychologists and behavioural economists have identified short-termist, impulsive System 1-style decision-making as a culprit in problems associated with overeating. These problems are intensified in obesogenic modern environments. Our metabolisms have evolved to suit a world in which food is scarce, but this is a mismatch with the abundance and easy availability of food stuffs, especially sugary treats, today. By collecting together in slimming clubs, we can overcome our instinctive impulses to overeat, and allow our System 2 thinking to dominate more easily. In ensuring this outcome, clubs work much better when a leader takes responsibility for motivating and coordinating their followers.

Clubs are not all about unadulterated self-interest, however. The Environmental Champions study mentioned above in the context of pro-environmental workplace behaviours also explored methods to improve environmental decision-making at home, using club-like groups as a forum to encourage pro-social behaviours. The researchers brought householders together via an 'EcoTeams' programme to look at common household habits and behaviours. At neighbourhood meetings, the Environmental Champion leaders briefed their local communities about better practices for energy use. As for the Environmental Champions scheme, the positive impacts of EcoTeams were significant: 16 per cent of the households involved went on to adopt green energy tariffs, 37 per cent installed energy-efficient light bulbs and 17 per cent reduced their domestic heating consumption. Whilst there was a degree of self-selection involved (people who were already environmentally aware were more likely to join), nonetheless the participants stated that the EcoTeams programme worked for them because it was focused on imparting and communicating practical knowledge via teamwork and collaboration.[22]

Congregations

Less objective and more subjective influences are crucial when questions of spirituality and identity enter the mix, for example in congregations. The comedian Danny Wallace's non-religious Join Me movement was a good example. Wallace formed his congregation by putting an advert in *Loot* magazine inviting people to join his movement. A surprisingly large number of people signed up, even though they did not really know what Join Me was about.[23] Within a congregation, the goals of the group gathered together seem genuinely constructive in terms of building a community, even from the perspective of an outsider. Wallace argued that, in the case of Join Me, the risk-taking inherent in joining a group of strangers was also attractive. The positive outcome was membership of a welcoming community with a common purpose.

In a religious context, the relationship between leader and followers in a congregation represents a mix of the objective rewards from joining a group alongside a sense of belonging and purpose, linking to some of the drivers of collective herding that we first explored in chapter 2. When it comes to religious congregations, at least from the perspective of believers, faith transcends the simple division between calm, deliberative System 2 and instinctive, emotional System 1 thinking. Strongly held beliefs are not obviously objective but nor do they seem to satisfy any basic needs or instinct. Understanding the System 1 dimensions returns us to some of the Freudian insights about our unconscious motivations. We are captured by ineffable and transcendent beliefs, operating beyond either reason or instinct – whether they be about a belief in a God, gods or other spiritual beings, or a belief that there is no God at all. Religious feelings puzzled Freud. He struggled both to find religious sentiment in himself and to categorise religious feelings more generally. He recalls his correspondence with an unnamed friend, who wrote to him about religious sentiment and the way in which religious leaders can take hold of it:

> [My friend] was sorry I had not properly appreciated the true source of religious sentiments ... [It] consists in a peculiar feeling ... a sensation of 'eternity', a feeling as of something limitless, unbounded – as it were, 'oceanic'. This feeling ... is a purely subjective fact, not an article of faith; it brings with it no assurance of personal immortality, but it is the source of religious energy which is seized upon by various Churches and religious systems, directed by them into particular channels, and doubtless also exhausted by them. One may ... rightly call oneself religious on the ground of this oceanic feeling alone, even if one rejects every belief and every illusion.[24]

We cannot easily understand religious and spiritual congregations as a product of self-interest, or by applying logic and analysis, but neuroscientists are starting to explore what drives different religious beliefs. A team of US neuroscientists used fMRI to scan the brains of fifteen committed Christians and fifteen nonbelievers asked to think about a range of religious propositions (the Virgin Birth, God and so on) and non-religious propositions. They found that emotional and reward-processing areas of the brain, as well as areas associated with cognitive conflict, were engaged more strongly by religious thinking, while thinking about non-religious facts engaged areas of the prefrontal cortex associated with memory retrieval.[25] This evidence suggests that religious beliefs are more likely than non-religious beliefs to reflect instincts and emotions, but logic and reason have a role too. Overall, religious congregations are unified by interplays between System 1 and System 2 thinking.

Ultimately, joining congregations and other groups gives many people a sense of an existence that is beyond the individual, and even beyond the groups themselves. By joining a religious congregation believers can feel connected with a faith community stretching across the world, and religious leaders play a key role in promulgating the message. In congregations, the hierarchy separating leader from follower is less clear than in secular contexts because ultimately the congregation is led by a spiritual goal and/or a belief in some higher being. As Gustave Le Bon observed in his description of the psychological crowd,

> [such] crowds are about the realm of sentiment ... in the case of every thing that belongs to the realm of sentiment – religion, politics, morality, the affections and antipathies ... [in the crowd] the most eminent

> men seldom surpass the standard of the most ordinary individuals . . .[26]

Cults

Unlike congregations, which are often driven by benign purposes, cults illustrate some of the most perverse aspects of the symbiosis between copycat followers and contrarian leaders. In everyday language, the word 'cult' is often used in a pejorative sense – though whether we believe a religious organisation is a cult or a genuine religion is a matter of subjective opinion. One feature of cults that distinguishes them from conventional religions is that there is often a sinister relationship between the leader and his (rarely her) followers. The leader is perceived to be both mortal and divine, even though to an outsider he is just another human being.

We see this most starkly in the ancient world, when superstitions and beliefs had a powerful pull on ordinary people and there were stronger social hierarchies separating leaders from followers. One example is the ancient Egyptians' embracement of the 'cult of the living king'. During coronation rituals, pharaohs were accorded *ntr*, or godly status, via the union of their human self and the royal *ka*, or soul. When pharaohs' *ntr* and *ka* were united they became sons of gods. Subsequent rituals reinforced this status, including ones in which the pharaoh would make offerings to his own deified self.[27] Illustrating the contrarian, maverick natures of many leaders, the pharaoh Akhenaten developed the cult of king-worship to new levels, with the monotheistic sun-worshipping cult of Aten the sun-disk at the centre.[28]

Akhenaten believed himself to be the son of Aten and encouraged Egyptians to worship him as the god's representative on Earth, with statues of Aten replaced by images

Figure 8. Leading the cult of Aten: Akenhaten worshipping the sun-disk.

of Akhenaten and Queen Nefertiti, his wife. Akhenaten mandated himself as a god to replace Egyptians' traditional polytheistic worship of the gods. He closed temples, eradicated priests and removed all references to old gods from places of worship and monuments. Akhenaten was authoritarian, and exerted his leadership role dictatorially. Ordinary Egyptians suffered great hardship and short life expectancies, and many whose names referred to other gods were obliged to change them. Whilst Akhenaten's reign was relatively short-lived – he probably reigned for just seventeen years or so – the historical significance of his cult continues today: it was the first known monotheistic religion.

Jim Jones' Peoples Temple, introduced in chapter 2, is a modern example of a sinister cult with a charismatic leader. Many would claim that Scientology is another. Its figurehead is a single hypnotic leader, David Miscavige, and it uses the cult of celebrity to build its profile through the prominent endorsements of the likes of Tom Cruise, John Travolta and Kirstie Alley. In common with religious cults, it veils itself in secrecy and exclusivity, drawing on an untestable mythology about its origins. While Scientology claims a basis in science and psychological evidence, to outsiders it seems to mostly be a product of an imbalance between emotion and deliberation – with emotion and instinct operating unilaterally, and without the moderating influence of reason.[29]

Overall, if clubs are about the dominance of System 2 thinking, and congregations are a balance between System 1 and System 2 thinking, cults are much more about a System 1 emotional response. Cult leaders exploit their followers' insecurities by encouraging them to sever their ties with friends, families and communities.[30] Cult followers are seeking comfort and reassurance in the face of fear and uncertainty, a response we have evolved to help us cope with the stresses of life both large and small. For cult leaders, their followers are essential to their power and existence. Without followers, the cult would not exist.

Modern idolatry

The faith placed in a cult leader often leads to terrible outcomes. A more benign version is the hero worship common in our everyday lives, manifested in fans' adulation of stars of stage and screen. This again is a leader–follower symbiotic relationship. The incentives for the leaders – the pop stars and teen idols – are clear: they accumulate money, fame and glory from their fans' attention. But many of us cannot fathom the reasons for fans' adulation. Beatlemania is a classic example of hero worship. It emerged in the UK in 1963 and reached an apex in 1964 when around 73 million viewers watched The Beatles' performance on *The Ed Sullivan Show*. In person, fans exhibited manic, hysterical behaviour – screaming, swooning and throwing their knickers at the group.[31] In his book *Beatlemania*, the journalist Martin Creasy writes that, at one concert, fifty collapsing girls were carried out within five minutes, sobbing hysterically. At another concert in Glasgow, groups of over 3,000 fans got caught in a frenzy, colliding with each other in the melee.[32]

These episodes of group mania and collective herding have much to do with the nature of the group, herd or mob as an entity in itself – but what is going on specifically in the relationship between the star and their fans and groupies? Many observe that fandom is a form of pathology – perhaps fans are exhibiting some form of mental illness. The reality is likely to be much more complex, however, and in any case, such diagnoses of mental illness fail to capture anything about the actual relationship between the fan and the star. The star offers something to the fan. They are a symbol of something attainably good and desirable.

Fan hysteria is not a new phenomenon, and fans are not always female. Nor do fan riots require modern media to start and sustain them. Beatlemania-like frenzies were observed by German writer Heinrich Heine in 1844. Heine wrote about the craze for the composer Franz Liszt that swept through Europe after Liszt's compositions began to attract a lot of attention around Germany in 1841. Lisztomania triggered episodes of hysteria amongst the composer's growing fan base. After one of his concerts in Berlin, fans mobbed him, fighting over his clothes and jewellery.[33] Extreme emotional responses have also been observed in the admiration of art. The quickening heartbeats, fainting and hallucinations experienced by some gallery-goers standing before particular pieces – identified as 'Stendhal syndrome' by researchers – are symptoms not dissimilar from those experienced by fans in the presence of their idols.[34] Overall, fans' worship of their idols is more than just a form of psychopathology. Nonetheless, and in common with cult members, System 1 emotion and instinct is dominating their System 2 reason and deliberation. These responses are magnified by external institutions, including markets. Businesses selling merchandise can make a lot of money by encouraging and amplifying fans' hysteria. Fandom is also manipulated by stars' managers. During their US tour, The Beatles sold not only millions of records but made over $2.5 million in revenue from selling branded merchandise. For modern stars, the rewards are even greater.

Political tribalism

In the secular world, tribes are the corollaries of cults. Tribalism has been an enduring feature of human interactions ever since we lived in hunter-gatherer communities. It is also another manifestation of the complex relationships between leaders and followers. The primitive impulses to join a tribe are seen in the modern world. In modern democracies, tribalism manifests itself in our political relationships. Political leaders, often in cahoots with business leaders, can distort voting patterns and exploit crowds by the manipulation of information.

John Maynard Keynes observed that diverting a thirst for power into more material ends might be beneficial for society:

better that this world is full of ruthless robber barons than brutal dictators:

> dangerous human proclivities can be canalised into comparatively harmless channels by the existence of opportunities for money-making and private wealth, which, if they cannot be satisfied in this way, may find their outlet in cruelty, the reckless pursuit of personal power and authority, and other forms of self-aggrandisement. It is better that a man should tyrannise over his bank balance than over his fellow citizens; and whilst the former is sometimes denounced as being but a means to the latter, sometimes at least it is an alternative.[35]

Keynes' use of the word 'sometimes' is telling: in our modern world, the relationships between political leadership and commercial interests can be worryingly close. We do not benefit from the simple separation of markets from politics as advocated by Keynes. Mass media have enabled a convergence of politics and business. Just because someone diverts their activities into business rather than political leadership does not prevent them from wielding excessive power and influence on a global scale – especially if they control the media. Donald Trump is one example; another is press baron Rupert Murdoch, whose enormous business empire enabled him to wield considerable international political power too.

Modern political tribalism is intensified by the ways in which we can now herd together, facilitated by social media. Social media allow a much stronger relationship to develop between leaders and followers. Facebook and Twitter are direct conduits for personal information, which increases a sense of intimacy. Twitter feeds, Facebook walls and other online forums mean that today's followers feel a disproportionate sense of connection to, and responsiveness from, their leaders, even though most will know they are conversing with social media teams propounding focus-grouped messages. Nonetheless, these social media tools give followers the impression that they have a tangible relationship with their leaders, consolidating the feeling that they are bound together with them and other like-minded followers by common goals as well as a shared identity. Thus, in the run-up to the UK's EU membership referendum, the many pictures and videos of then leader of UKIP Nigel Farage drinking beer in a pub circulating on social (and mainstream) media increased his support. In portraying him as an 'ordinary bloke' the images directly appealed to his supporters' sense of identity – even though, in reality and unlike the vast majority of UKIP supporters, Farage comes from a privileged and affluent background.

Political herding: reason versus emotion

We do not always reason carefully through all the facts when we make our political choices. But that does not mean that reason plays no role at all. Sushil Bikhchandani and his colleagues have applied their concept of information cascades, as explored in chapter 1, to American political campaigns. Voters balance their private information about the different candidates against the social information they can gather about other voters' likely choices. When reliable information about the different candidates is not easy to find then social information will dominate, tipping undecided voters into joining the herd.[36]

Convention also dictates the strategies of candidates on the ballot papers. As noted in the previous chapter, the *median voter theorem* suggests that, for the average politician, it makes sense to identify the average position on a given issue, and then to build a political manifesto accordingly. Again, this connects with risk, because the average politician, lacking much independent conviction, will gravitate towards a conformist position. That is the least risky strategy if they want to be elected. Political times are changing, however, and social media are shifting the centre of gravity away from the average. In November 2015, *The Economist* presented some evidence to show that strongly right-wing and strongly left-wing parties have a relatively substantial social media presence, perhaps because social media reward stark sound-bites ahead of subtle messages.[37] This evidence taps into the idea that taking risks can deliver rewards. Blunt statements risk easy condemnation, whereas nuanced communications leave the messenger with more leeway for interpretation. Historically, political extremists found it difficult to take these risks because it was not easy to promote an extreme position via traditional media. With Twitter, Facebook and other forms of social media, this constraint has disappeared.

In voters' adoration for their political leaders and in political decision-making more generally, emotions are everywhere, reflecting System 1 thinking. Daniel Kahneman himself noted the dominance of emotions in the run-up to the UK's 2016 referendum to leave the European Union – presciently worrying that destructive psychology was blinding people to the long-term consequences of Brexit. In an interview for the UK's *Daily Telegraph* published just a couple of weeks before the referendum, Kahneman observed, 'The major impression one gets observing the debate is that the reasons for exit are clearly emotional . . . The arguments look odd: they look short-term and based on irritation and anger.'[38] This reliance on emotion is all but inevitable. Voters don't have time, and sometimes lack the expertise, to research and understand all the details of the policies put forward by electoral candidates and lobby groups, let alone to examine the minutiae of politicians' backgrounds. Added to these constraints is the fact that political news has become so noisy and unreliable that even those who do have the time and expertise to interpret it all are still left feeling confused. It is easier, and in some ways more satisfying, to fall back on System 1 thinking.

Voters may also be using political herding heuristics in a relatively unemotional way. As we have seen in previous chapters, we use heuristics – simple rules of thumb – to help us to make quick decisions. In the case of political herding, each individual voter knows that his or her single vote is not going to change the outcome of any given election – so there is no incentive to be fully and completely informed about the options. We don't have to spend a lot of time thinking deeply about our political choices because individual voters do not have to take responsibility for aggregate outcomes. This creates a free-rider problem. No-one is properly incentivised to make an effort in searching for facts. A diffused sense of responsibility for the outcome encourages individuals to express themselves via an individual protest vote, say for an extreme candidate or outcome.

This situation is exacerbated in an uncertain world when information is muddy. When we struggle to assess the trustworthiness of information, it impairs our ability to balance different information sources against each other. The Brexit vote illustrates some of the problems that can emerge when voters don't trust the information promulgated by their leaders. In the run-up to the referendum, both Leave and Remain circulated misleading information, creating widespread confusion. Ordinary voters could not know who was being more truthful and whom they could trust. There was no verifiably trustworthy group to follow.

Herding heuristics and social learning strategies only work well when we can assume that we are not being manipulated. The way in which we learn from others is complicated by the emergence of 'fake news' – defined by economists Hunt Allcott and Matthew Gentzkow as news stories that are verifiably false and intentionally devised to mislead readers. Drawing on data collected during the 2016 US presidential election, Allcott and Gentzkow used econometric tools to analyse large numbers of fake news stories. They conducted a post-election survey of nearly 11,000 American voters to estimate how many articles their respondents had seen, and identified twenty-one fake news stories which had been repeatedly read and remembered over the course of the campaign. Their evidence suggests that fake news stories were worryingly influential in the election outcome.[39] If we herd behind others on the basis of false 'information' deliberately circulated by politicians and their spin doctors, then we are being manipulated without even knowing it, especially when fact-checking is difficult. When this fake news is psychologically and emotionally laden, emotional influences creep in without us realising, making us even more vulnerable to manipulation.

The leader of the 'Free World'

Risk-taking is an essential ingredient for political success. US President Donald Trump's success is a story of a political triumph based around a business-world entrepreneurial risk-taking strategy adapted to the political domain. By taking what others might have thought were extreme risks, he was able to reap large rewards. Alongside his own risk-taking strategies, he also manipulated the conformist tendencies of his in-group. His battle against Hillary Clinton in the 2016 US presidential election illustrates some of the tribal and political tensions, and their links with relationships between leaders and followers. Both Trump and Clinton were controversial candidates. Both were rich members of the elite, Trump in business and Clinton in politics. Both had been involved in damning controversies but, on the surface at least, Clinton had the extensive political experience, a distinguished academic and professional background and evidence of real intelligence (though she may also have faced the additional obstacle of voters' bias against women).

Trump leveraged the wealth accumulated from his business empire to finance his ultimately successful election campaign, succeeding despite lurid and compromising allegations swirling around him – allegations that perhaps would have scuppered his aspirations in more stable times. For Trump and his supporters, his victory was marvellous and spectacular, not only because few pundits were able to predict the outcome. For others, it was a disaster, and seemed to unleash increasing division, polarisation and tribalism, not only within the American electorate but also across the globe.

How was Trump able to attract such massive electoral support? He seems to share little in common with those who voted for him, whether blue-collar or white-collar workers. Trump was heir to an enormous fortune and has lived a life of wealth and privilege, and yet ordinary Americans, some of whom eke out a living in the most straitened of circumstances, believe that he is their champion. Trump was innately able to encourage ordinary people to identify with his rebellion against established elites. His emotive and impulsive outbursts were shocking to some, but very appealing to others. Through the election cycle and after his inauguration as president, he seemed unafraid of conflict. He took on all comers, especially those conventionally regarded as authoritative – including key members of his own Republican Party as well as American intelligence advisers and the judicial system. Trump's calls to rebellion and his eventual victory allowed ordinary people to believe that they too could wrest control of their own destinies away from the political elite.

The ability to tap into voters' sense of identity links back to psychologist Henri Tajfel's insights. As we saw in chapter 2, Tajfel explored how easy it is to build identity with our in-groups and encourage conflicts with our out-groups. By encouraging fear of the out-group, the in-group creates a strong identity that feeds on itself, as members of the group reinforce each other's views, strengthening the group's power. This is the basis for Trump's populism. Trump exploited fear of the out-group to a controversial degree – such as during the election campaign, when he accused some Mexican immigrants of being rapists and criminals, and pledged to build a wall along the US's southern border. On a platform of 'making America great again' he tapped into Islamophobia and fears about terrorism with his divisive and seemingly ill-judged 'Muslim ban', enforcing stricter rules on visas for travellers from a selected group of Muslim-majority countries (tellingly, not including Saudi Arabia).[40] So it may be surprising to think that, possibly without knowing it, Trump was displaying a wily social intelligence. He understands crowds and what motivates them – an innate talent perhaps, but consolidated during his time as a TV celebrity on the US version of *The Apprentice*. His supporters are simultaneously copycats and contrarians, herding together with a minority of other Trump-supporting copycats while rebelling against the majority of voters who oppose him. Trump did not need majority support to legitimise his role as leader, and his plummeting approval ratings after the election have not made much practical difference either. Within a year, Trump's remaining support base had fallen to around 35 per cent, but still large enough, and perhaps more importantly, fanatical and cohesive enough, to give him a solid base of power. Trump's cunning came in understanding that his minority tribal following was more interested in powerful social media messages than demonstrable facts. His accusations of 'fake news' became notorious and he attacked his detractors on a near weekly basis, circulating other inflammatory statements via Twitter every day. His advisers, and websites associated with him during his election run, orchestrated highly effective campaigns using spoof and smear stories. Trump realised that an ability to understand and tell the literal truth is not the path to political power.

So why would he tell the truth? PolitiFact monitors the verifiability of politicians' statements in the US, and their verdict on Donald Trump's presidential campaign was that 70 per cent of his statements were in the range 'mostly false' to 'pants on fire', with another 14 per cent only 'half true' – so only 16 per cent of his statements can be said to be true.[41] By contrast, during her presidential campaign, Hillary Clinton's record was 26 per cent in the 'false' range with 51 per cent seeming to be 'true' and 'mostly true', suggesting that most of what Clinton said was checking out.[42] Quite aside from the other reasons not to vote for him, the questionable veracity of Trump's statements did not seem to deter a large number of voters. Why wouldn't a ruthlessly ambitious politician lie if there is no institutionalised penalty for manipulative grandstanding? And if politicians' assertions appear in a Twitter feed, ephemeral and quickly removed, then they may even escape much in the way of rebuke – though the message may still have an emotional impact on supporters.

So, whilst it might be easy to conclude that those voting for Trump were not as well informed as those voting for Clinton, this reflects a poor and potentially divisive understanding of the dynamics between leaders and followers. Democracy is built on principles of consensus. When facts are hard to find and the world is uncertain and confusing, then consensus is built on unedifying foundations. Populist politicians have encouraged us to reject objective information and the judgements of experts, and their ability to promulgate their populist messages quickly is amplified by Twitter and other social media. The influence of social media was a key factor behind the seismic political changes of 2016, not only the election of Donald Trump but also the UK vote for Brexit. Indeed, as US Senate committees continue to deliberate on whether social media were exploited by Russian interests keen to deliver a victory for Donald Trump in the 2016 presidential election, the platforms are under fire. Far from being heralded as a channel for triumphant democracy during the Arab Spring uprisings of 2010–12, social media are now castigated as providing the means for the exertion of sinister geopolitical manipulations at the highest echelons of international power.[43] Whatever the outcome of the investigations, it's clear that Trump is a genius of reinvention and, curiously for such a largely self-absorbed demagogue, has an acute kind of social consciousness. If he did not, then he would not have been able to inspire such a loyal herd of copycats to follow his lead.[44]

Exacerbating the lack of trust in news and the dubious influences of social media is the sheer volume of information at our disposal. Living in our interconnected, indeed overconnected online world, we are exposed to a relentless, inexhaustible feed of information coming very quickly from lots of different sources. As recent studies have shown, when large volumes of information are contradictory and confusing, people struggle to distinguish between spoof and real news stories.[45] How do we know what is good information and what is bad? How do we extrapolate information from the noise surrounding it? Social media have their virtues, but they cloud information. Twitter trends constantly shift, and within a minute we can accumulate dozens of tweets conveying different pieces of news or opinion – and it is often, and increasingly, hard to differentiate one from the other. Under these conditions social media have a lot of power, most effective when they tap directly into our System 1 quick thinking processes, so we will often process this confusing volume of information unconsciously.

It therefore makes sense for voters to decide on grounds not directly related to objective facts, because objective facts are neither available nor reliable. Like the restaurant-goers choosing between restaurants, voters, when they have very little reliable information to draw on, tend to follow a herd of like-minded people and/or a persuasive leader with whom they most readily identify. Or they rely on information from those closest to them – the echo chambers, in which people's views and opinions are reinforced by those who already agree with them. Social media magnify this effect.[46] On Facebook, Twitter and other platforms we tend to read the posts of those in our family and friendship groups who are like us and who we like. Our views are further clouded by confirmation bias. We circle around our own opinions – following those whom we have selected to follow, often because we already agree with them.[47] Similarly, we can easily fill our screens with preferred media outlets, whether BuzzFeed, Breitbart or Reddit, which match our existing views of the world, confirming our prejudices with large volumes of information day and night. Mistrust, confirmation bias and social media collide, leaving us in politically dangerous situations.

We might say that we want honesty from our political leaders, in the same way we want our scientific experts, doctors and lawyers to be honest; but ultimately perhaps we want our political leaders to represent us and our beliefs. In this our System 2 logical, deliberative assessment of facts is less important than the System 1 emotional, identity-focused impact that our leaders have on us. Our political decisions are not dictated by a reasoning search for the truth.

The dominance of subjective over objective influences in our political choices means that we are less concerned about how honest our politicians are about specific facts, and more concerned with the convictions that they communicate.[48]

In today's 'post-truth' political era, the clever shaping of politicians' public profiles to tap into our quick, emotional System 1 instinctive decision-making is effective in manipulating our choice of leaders. Fake news suits the System 1 thinking style well. It is designed to be digestible – usually consisting of simple, emotive messages that we can easily process using quick decision-making heuristics. When we have little trustworthy information to engage our more logical and deliberative System 2 thinking, it is not surprising that System 1 thinking dominates and sways the crowd's political opinions. Populist politicians build support via an appeal to System 1 emotions and instincts, and social media are a very effective conduit for these. Their emotive messages capture our imagination and connect with our identities much more immediately than any information we might gather from trawling through manifestos or unpicking the finer points of political policy changes.

We have seen in previous chapters that we are copycats in many aspects of our everyday lives. Whilst contrarians are (obviously) a minority group, copycat followers and contrarian leaders are often mutually dependent. Copycats joining together in crowds and herds need contrarian leaders to lead them together in one direction. But, perhaps less obviously, contrarian leaders need copycats too. Leaders cannot be leaders without followers. As we have seen in this chapter, our politicians have a talent for encouraging political tribalism and they rely on copycats for their success in building their political tribes. Today's social media platforms mean that political leaders can build this tribalism in myriad ways. Social media have also empowered individuals who would be excluded from the political process in previous eras.

Overall, are we *always* contrarian leaders or copycat followers? It is likely that our choice, to the extent that we have one, will depend on the context in which we find ourselves. Our different inclinations to follow or lead will also be driven by a combination of System 1 emotion and instinct and System 2 logic and deliberation. This is a crucial insight more generally, as we have seen throughout this book. A delicate balance between System 1 emotion and System 2 reason not only propels our decisions to copy or rebel, it also determines the copycat and contrarian roles we choose.

Conclusion

COPYCATS VERSUS CONTRARIANS

In this book, we have traced a path from economics through psychology and sociology to neuroscience, behavioural ecology and evolutionary biology. We have explored what drives copycats and contrarians from a range of different perspectives. Mostly our copycat natures dominate, encouraging us to herd. Sometimes we herd purely out of self-interest in a clever and analytical way. At other times our herding is more a collective phenomenon, driven by instincts and emotions. Sometimes herding is a mixture of these different influences.

The question we have yet to answer is whether or not it is good to be living in a world so dominated by copycats herding together and following leaders. In a primitive world, our strong tendencies to copy and follow probably served us well enough. Our antediluvian instincts evolved to help us survive in harsh natural environments, not just to ensure our survival as individuals but also reflecting evolutionary pressures to ensure survival of groups and tribes, genes and our species as a whole. Whether these herding tendencies work well today is much less clear. We might think that our daily lives are easy compared with those of our hunter-gatherer ancestors. Resources are relatively plentiful. Information and connections move and develop rapidly via modern technologies. We can easily build social relationships with people we have never met, and yet our conformist instincts can play a powerfully destructive role.

But there is a dark side to modern progress. Whilst modern technologies may seem to have enabled substantial improvements in our standards of living, they have also hijacked our old evolved survival strategies. Conflicts between our modern selves and our evolved selves are made more destructive when our evolved instincts to copy each other are rapidly channelled via modern technologies uninfluenced by the personal social sanctions and limits that small groups can and did impose when we lived in more concentrated communities. In our modern, computerised, globalised and deeply interconnected world – in which money, information and expectations move so fast – herds can build enormous momentum which is difficult to monitor and control. The pace of technical innovation and the changes to our artificial environments have been much too fast relative to evolutionary timescales and we have not had time to evolve new forms of adaptive advantage. Are we really fit to survive in a globalised world in which our tendencies to herd and conform are magnified by all the high-speed technologies that human ingenuity has invented? Perhaps not.

If we take a more critical look at the impacts of specific modern technologies, we might notice that some technologies have made our lives harder, not easier. There is no doubt that some technologies have made enormous, positive contributions. Medical advancements in particular have transformed both the length and quality of our lives – and a Luddite approach is not going to solve problems. What we do need to think about is how new technologies have disrupted the equilibrium between individual and social interests. Today, these are not as easily aligned as they were at the dawn of civilisation. What is best for the individual is diverging further and further away from what is desirable for economies and societies as a whole. The case for unfettered markets is less clear in a world of computers, big data and social media. Our evolved herding and copying instincts have enabled the growth of inefficient, counterproductive and, at worst, destructive forms of behaviour that could not have emerged in primitive settings in which the number and range of person-to-person connections was limited in reach and complexity.

Social media have had a particularly destructive influence, hints of which have appeared in all the previous chapters. Social media are conduits for fake news and false information, and this disrupts even the most rational social learning processes associated with System 2 self-interested herding. Social media tap effectively into quick System 1 thinking and the emotive, impulsive forms of collective herding, at the same time disrupting the balance between collective and self-interested herding.

Whilst in many senses social media have helped us to build our stores of knowledge and understanding, these platforms can also disempower contrarians, and in this way we are losing a richness and diversity of information and opinions. Conventional views are cooked up, reinforced and replicated in the echo chambers of Twitter, Facebook and other news-sharing sites. Online 'town hall' conversations shut down controversial or contrary views – like 'no-platforming' speakers in university debates. Balancing controversial opinions is tricky. We have good reasons to curtail immoral and unethical opinions, but the boundary between what is unethical and offensive to almost everyone and what is offensive just to a specific group is fuzzy. Contrarians are hounded by

Twitter trolls. The polarised debates around Brexit are an example of this. Whether a 'Remainer' voting to stay in the EU or a 'Brexiteer' keen to leave, expressing opinions about the Brexit vote catalysed vicious reactions from the opposing group. Social media distort the dissemination of expert insights, such as those to do with medical and scientific breakthroughs. The sheer volume of noisy, contradictory information distributed via these platforms means that it is difficult to judge the evidence effectively, even for someone who might aspire to be completely logical and objective. Social media give leaders another weapon to use in manipulating and controlling their followers. Leaders can tap into the herding instincts of their copycat followers to manipulate their choices, with wide negative consequences – for instance, in promoting political tribalism, as we saw in the last chapter.

Fragile, unstable and unreasoning attitudes towards experts, elitists and migrants, and the rise of extreme political positions, are all partly formed by strong instincts to follow public opinion in an emotive way rather than focusing on the facts presented. Social media almost appear to be custom-built to serve this quick, instinctive and unreasoning behaviour. Citizens' confusion and mistrust about information and news reflects the fact that, in the modern, 'post-truth' social media age, the usual news outlets have been replaced by information conduits that are not confined within the bounds of traditional journalism's fact-checking protocols. Without reliable information sources, copycats can be led by their contrarian leaders down paths that from the outset they neither understood nor anticipated.

Within our social networks, too, our conformist copycat tendencies have been distorted by modern technology. Social networks have grown and changed rapidly with computers and the internet, creating an overconnectedness in the modern world which most of us probably don't think about too carefully anymore. Before the internet, social networks were largely constructed around the social bonds people had with others close to them, whether relatives, work colleagues or neighbours. Social theorists explain these social networks as a form of social capital built up from our social investments in relationships with others around us.[1] These ties are hardened by social norms that evolve alongside our social networks, and often these social norms are rigid, inflexible and resistant to change. They can operate and develop in a diffuse way over long periods of time, for example in the evolution of class hierarchies and social stratifications such as the Indian caste system. Our sense of identity parallels the strength of the ties originating within the in-groups that form part of our social networks.[2]

We form weak ties with others in our professional networks and associations via online social platforms such as LinkedIn, ResearchGate and Academia. In these networks, we are forming social bridges with other people and different groups – allowing us to make connections we might not otherwise make. These online networks can be useful and productive in a general way, for both the individual and the group. They enable us to exchange ideas quickly, to build our professional relationships and identify new employment or business opportunities. We certainly do not want to return to the rigid hierarchies associated with traditional, discriminatory social network structures. But the view, sometimes propounded by social theorists, that strong ties and bonds between us are bad and weak ties are good is harder to defend when social networks and social media collide. Strong ties and bonds help in-groups to build strength and power. Gangs, for example, are characterised by the strength of the relationships and loyalty between their members. As we saw in chapter 2, these strong ties can have destructive impacts in terms of violence and discrimination against out-groups, in some cases to the extent that we are prepared to put our in-group at a disadvantage in our conflicts with our out-groups.[3]

But in a technologically dominated world, we should be worrying about weak ties and bridges too. Online, weak ties are just as likely to promote discrimination and negative attitudes and behaviours – Twitter trolls and cyber-bullying, for instance. The innumerable weak ties we develop via social media and the overconnectedness enabled by those platforms also have other negative consequences for our well-being. They mean that businesses can easily invade our privacy and exploit our willingness to share information in the process of impressing others. We can never properly switch off from work when our work email is only a smartphone bleep away. As employees, we suffer the consequences of increased stress and the inability to relax, but our employers suffer too if that erodes our productivity. Online social networks encourage copycats' obsession with what everyone else is thinking, and at the same time enable the construction of impossibly rosy online profiles. If people only ever look at everyone else's very best, filtered sides on Facebook and Instagram it's no surprise that confidence and self-esteem are far harder for young people to find today than they were thirty years ago. Rising teenage suicide rates are some of the saddest consequences of our shift into the copycat-dominated online world.

Taming copycats and contrarians

So: what should we do? What policy tools will work best to tame our herding and anti-herding instincts when they are destructive, or leverage them when they are beneficial? In this book we have explored some of the ways in which our herding instincts can be used to encourage people to follow their neighbours in more constructive behaviours – in the context of energy decision-making and sanitation habits, for example. Using our conformist inclinations as a policy tool for social 'nudging' has become very popular. Small changes in the way information and options are presented encourage people to change their choices in a more constructive direction.[4] Social nudges are now used extensively by behavioural public policymakers, such as the UK's Behavioural Insights Team and its spin-offs.[5] But given the many ways our copycat natures are fallible, some of which we have explored in this book, perhaps policymakers should focus less on leveraging social conformity and more on controlling it, and/or encouraging mavericks and contrarians when these natures can help us onto a better path.

Using anti-herding as a policy tool is, however, a conceptual and logistical challenge. Almost by definition, it is difficult to manipulate the choices of anti-herding contrarians because their natures incline them to resist persuasion. Even so, at a time when we need mavericks to take risks, policy solutions can be designed around encouraging them. We may want to give more support to the experts espousing a contrarian view, assuming it is founded on good evidence. This is an idea that the American philosopher Michael Weisberg has explored. Weisberg found that when there are too few maverick experts relative to copycat experts, then landscapes of knowledge and new ideas are not fully explored. Too many copycats generate too few new ideas, slowing progress and innovation. We need to devise incentives for contrarianism. As we saw in chapter 7, Weisberg argues that there should be additional incentives for risk-taking in scientific research, and perhaps that insight should be extended to a wider range of occupations, including journalism and finance. Potential solutions include developing more rigorous standards for assessing the veracity of news stories or financial advice, and encouraging whistleblowing so that mistakes are identified and corrected quickly. At the same time, we need to ensure high professional standards and/or robust regulation so that gullible or ill-informed copycats, who might not have the expertise to judge the information – for example when digesting scientists' new research or esoteric economic insights from journalists, economists and financial advisers – are not exploited.

Another feature of modern life is the dominance of committees, but, as we have seen, committees can be hothouses of conformism and groupthink. Encouraging new social norms to encourage all committee members, not just the chairs and senior members, to express contrary opinions would help in reducing these tensions. Clear and transparent rules for deliberation on committees, as well robust guidelines for chairs of committees, might ameliorate some of the problems created by peer pressure, groupthink and copycats' tendencies to obey authority figures. Encouraging greater diversity on committees, so that different viewpoints are fully explored, could also be part of a solution. Another solution is to institutionalise roles for devil's advocates on committees, as is already the practice in the US defence and intelligence community.

We also need policies that effectively balance conformity and dissent.[6] To overcome the loss of private information incurred from our ingrained tendencies to follow others, one policy solution would be to ensure that better information and better education make us less dependent on others' opinions. For example, robust education and information campaigns could be introduced by impartial organisations, designed to help all voters understand the economic, political and legal institutions in which we live. Then, politicians would not be so easily able to hoodwink voters into believing unrealistic manifesto pledges, economic pronouncements and other political promises.

This book has explored the myriad ways in which our instincts, whether to imitate or rebel, affect our everyday lives. Is herding good for us as individuals? Is it good for society at large? Whether or not we decide that herding is desirable will depend on whether we take the perspective of the individual or society as a whole. Economic theory shows that herding often works well enough from the perspective of a self-interested copycat. Given market and institutional failures, individuals have rational reasons to collaborate, to look to the group, to copy and to herd – by observing and learning from watching others, by joining clubs and teams. From the perspective of groups and the human species, however, the benefits are less clear and will depend on context. The individual is sometimes dispensable to the group's interest. A blind instinct to join the group, to obey wrong-headed orders or to engage in acts of self-destruction such as self-mutilation, suicide bombing or self-sacrifice in wartime are all behaviours that prioritise one group over another, exacerbating inter-group tensions.

If we can develop a better understanding of the complex social interactions driving copycats and contrarians, then we will be better able to identify solutions to moderate herding and anti-herding when they are problematic, as well as to encourage herding and anti-herding when they are beneficial. But today's world is characterised by a potentially destructive imbalance. Our evolved natures, modern institutions, tribal politics, globalised markets and cutting-edge technologies have all allowed copycats and their leaders to thrive whilst contrarians and mavericks are marginalised. If we are to prevent a dystopian future dominated by groupthink, echo chambers, intolerance, inequality and conflict then we need to celebrate the best of what is unconventional, rebalancing our world so that copycats and contrarians can thrive together in tomorrow's world.

Endnotes

Introduction

1. Oliver Milman (2015), 'Cane Toad Sausages on Menu in Attempt to Save Kimberley's Northern Quolls', *Guardian*, 20 September. https://www.theguardian.com/australia-news/2015/sep/10/cane-toad-sausages-on-menu-in-attempt-to-save-kimberleys-northern-quolls; Angela Heathcote (2017), 'Toad Sausages are Saving Our Quolls', *Australian Geographic*, 31 August. http://www.australiangeographic.com.au/topics/wildlife/2017/08/toad-sausages-are-saving-our-quolls. See also Jonathan Webb, Sarah Legge, Katherine Tuft, Teigan Cremona and Caitlin Austin (2015), 'Can We Mitigate Cane Toad Impacts on Northern Quolls?', Charles Darwin University. http://www.nespnorthern.edu.au/wp-content/uploads/2015/10/4.1.35_can_we_mitigate_cane_toad_impacts_on_northern_quolls_-_final_report.pdf (accessed 30 September 2017).
2. The first use of the term 'anti-herding' I can find was by economists Canice Prendergast and Lars Stole, who used it to capture contrarians building reputation. See Canice Prendergast and Lars Stole (1996), 'Impetuous Youngsters and Jaded Old-Timers: Acquiring a Reputation for Learning', *Journal of Political Economy* 104(6), pp. 1105–34. Since then, the mentions of herding in the academic literature overwhelm the mentions of anti-herding by many orders of magnitude.
3. Liz Connor (2017), 'British People Display Amazing Queue Etiquette Without Being Told', *Evening Standard*, 2 May. https://www.standard.co.uk/lifestyle/london-life/british-people-display-amazing-queuing-etiquette-without-being-told-a3528366.html (accessed 5 September 2017).

1 Clever copying

1. Gary S. Becker (1976), *The Economic Approach to Human Behavior*, University of Chicago Press. These basic principles of rational choice have

also been popularised in general-audience books such as Steven E. Landsburg (1995/2012), *The Armchair Economist: Economics and Everyday Life*, New York: Free Press, and Steven D. Levitt and Stephen J. Dubner (2007), *Freakonomics: A Rogue Economist Explores the Hidden Side of Everything*, London: Penguin Books.

2. Gary S. Becker (1974), 'A Theory of Social Interactions', *Journal of Political Economy* 82(6), pp. 1063–93.
3. Neoclassical economics is the dominant, but controversial, paradigm in modern mainstream economics and is based around the idea that free markets are the best institutions for ensuring human welfare. Vilfredo Pareto is less well known, at least amongst mainstream economists, for his sociological insights; for example, see Vilfredo Pareto (1935), *The Mind and Society* [*Trattato di Sociologia Generale*], trans. Arthur Livingston, New York: Harcourt, Brace and Company.
4. Vilfredo Pareto (1906/1980), *Manual of Political Economy*, trans. A.S. Schweir, New York: Augustus M. Kelley. For a history of the term, see also J. Persky (1995), 'Retrospectives: The Ethology of *Homo economicus*', *Journal of Economic Perspectives* 9(2), pp. 221–31.
5. Adam Smith (1776), *An Inquiry in the Nature and Causes of the Wealth of Nations*, Book I, ch. II, pp. 26–7.
6. For a technical survey of economic theories of herding, see Christophe P. Chamley (2003), *Rational Herds: Economic Models of Social Learning*, Cambridge University Press.
7. Sushil Bikhchandani, David Hirshleifer and Ivo Welch (1992), 'A Theory of Fads, Fashion, Custom, and Cultural Change as Informational Cascades', *Journal of Political Economy* 100(5), pp. 992–1026; Sushil Bikhchandani, David Hirshleifer and Ivo Welch (1998), 'Learning from the Behavior of Others: Conformity, Fads, and Informational Cascades', *Journal of Economic Perspectives* 12(3), pp. 151–70.
8. Bikchandani, Hirshleifer and Welch (1992), p. 994.
9. Abhijit Banerjee (1992), 'A Simple Model of Herd Behavior', *Quarterly Journal of Economics* 107(3), pp. 797–817.
10. There are other explanations for crowded versus empty restaurants, built around economic assumptions of rationality, for example Gary S. Becker (1991), 'A Note on Restaurant Pricing and Other Examples of Social Influences on Price', *Journal of Political Economy* 99(5), pp. 1109–16. Essentially, Becker explains the crowded restaurant phenomenon in terms of the impact of one person's consumption on another's person's demand: one person choosing a restaurant will increase other people's demand for that restaurant, and so restaurant queues will grow.
11. Thomas Bayes (1763), 'An essay towards solving a problem in the doctrine of chances – communicated by Mr Price, in a letter to John Canton', *Philosophical Transactions of the Royal Society* 53, pp. 370–418.
12. This sort of evidence has an advantage over laboratory experiments because experimenters are not forcing experimental conditions on people. People can make their decisions naturally, and, because these decisions are not the product of experimental interventions, observed behaviours are more likely to be robust, and not an experimental artefact,
13. Arthur Fishman and Uri Gneezy (2011), 'A Field Study of Social Learning', Bar Ilan University Working Paper, 29 April. https://www.biu.

ac.il/soc/ec/fishman/wp/A%20field%20study%20of%20Social%20Learning.pdf (accessed 30 October 2017).

14. Charles Holt (2006), *Markets, Games and Strategic Behavior*, Boston, MA: Pearson/Addison-Wesley.
15. Lisa R. Anderson and Charles A. Holt (1996), 'Classroom Games: Information Cascades', *Journal of Economic Perspectives* 10(4), pp. 187–93; Lisa R. Anderson and Charles Holt (1997), 'Information Cascades in the Laboratory', *American Economic Review* 87(5), pp. 847–62.
16. Anderson and Holt do not explain why they think students would be capable of Bayesian reasoning. Did they think that the students knew Bayesian probability so well that they could easily and quickly apply it to the experiment? Probably not. It is more likely that the experimenters assumed that the students were acting *as if* they could do Bayesian calculations. Other economists describing similarly sophisticated decision-making often revert to the analogy used by Milton Friedman, one of the founding fathers of liberal economics, and his colleague Leonard Savage: expert billiard players are very good at manipulating the trajectories of balls around a billiard table; they play billiards *as if* they have a deep knowledge of the laws of mechanics, even though it is unlikely that they really do have such a sophisticated understanding. There are heated debates about whether this is a justifiable defence of unrealistic assumption, which we will not rehearse here. Suffice it to say that it is possible that our brains have evolved to have the capacity to make Bayesian calculations automatically without us having to think too hard about it. See Milton Friedman and Leonard J. Savage (1948), 'The Utility Analysis of Choices Involving Risk', *Journal of Political Economy* LVI, p. 298.
17. Economists Erik Eyster and Matthew Rabin have explored the idea that imitation and herding can be irrational because a rational agent will see that the beliefs of others are correlated, and will take this correlation into account. These rational agents will 'anti-imitate' some of those that they observe. Alternatively, they will realise that they should follow all those in the queue ahead of them as if they were one person, and essentially interpret their behaviour as one social signal, not many. See Erik Eyster and Matthew Rabin (2014), 'Extensive Imitation is Irrational and Harmful', *Quarterly Journal of Economics* 129(4), pp. 1861–98.
18. Banerjee (1992).
19. Other economists have analysed the problems of herding in an uncertain world. For example see Ignacio Monzón (2017), 'Aggregate Uncertainty Can Lead to Incorrect Herds', *American Economic Journal: Microeconomics* 9(2), pp. 295–314.
20. For introductions to the game theory literature see Ken Binmore (2007), *Game Theory: A Very Short Introduction*, Oxford University Press, and David M. Kreps (1999), *Game Theory and Economic Modelling*, Oxford University Press. The seminal but esoteric economic text is John von Neumann and Oskar Morgenstern (1944), *The Theory of Games and Economic Behavior*, Princeton University Press.
21. Cited in Brian Skyrms and U.C. Irvine (2001), 'The Stag Hunt', *Proceedings and Addresses of the American Philosophical Association* 75(2), pp. 31–41.

22. See also Desmond Morris on social signalling as part of our search for meaning (linking to identity and a need to belong); for example Desmond Morris (1969/1994), *The Human Zoo*, New York: Vintage.
23. George A. Akerlof and Rachel E. Kranton (2000), 'Economics and Identity', *Quarterly Journal of Economics* 115(3), pp. 715–53. See also George A. Akerlof and Rachel E. Kranton (2011), *Identity Economics: How Our Identities Shape Our Work, Wages, and Well-Being*, Princeton University Press.
24. Henry Farrell (2015), 'With Your Tattoos and Topknots, Who Do You Think You Are?', *Washington Post*, 28 July. https://www.washingtonpost.com/news/monkey-cage/wp/2015/07/28/with-your-tattoos-and-topknots-who-do-you-think-you-are/ (accessed 7 September 2017).
25. Diego Gambetta (2009), *Codes of the Underworld: How Criminals Communicate*, Princeton University Press.
26. For more on virtue signalling, see James Bartholomew (2015), 'The Awful Rise of "Virtue Signalling"', *Spectator*, 18 April. https://www.spectator.co.uk/2015/04/hating-the-daily-mail-is-a-substitute-for-doing-good/ (accessed 7 September 2017).
27. Naimil Shah (2016), 'Why Poor People Buy TVs', *Medium*, 22 October. https://medium.com/@naimilshah/why-poor-people-buy-televisions (accessed 14 September 2017).
28. See also the economist Harvey Leibenstein on how self-interested herding emerges in the context of consumer choice: Harvey Leibenstein (1950), 'Bandwagon, Snob, and Veblen Effects in the Theory of Consumers' Demand', *Quarterly Journal of Economics* 64(2), pp. 183–207.
29. B. Douglas Bernheim (1994), 'A Theory of Conformity', *Journal of Political Economy* 102(5), pp. 841–77.
30. John Maynard Keynes (1936), *The General Theory of Employment, Interest and Money*, London: Macmillan and the Royal Economic Society, p. 158.
31. O. Johansson-Stenman and J. Konow (2010), 'Fair Air: Distributive Justice and Environmental Economics', *Environmental Resource Economics* 46(2), pp. 147–66, and G. Brown and D.A. Hagen (2010), 'Behavioral Economics and the Environment', *Environmental Resource Economics* 46(2), pp. 139–46.
32. 'Shot? The National Rifle Association', *The Economist Espresso*, 28 February 2018. https://espresso.economist.com/2e7fc7cb9bf8baacf29f1b7286976f53 (accessed 6 March 2018).
33. D. Kahneman, J.L. Knetsch and R.H. Thaler (1986), 'Anomalies: The Endowment Effect, Loss Aversion, and Status Quo Bias', *American Economic Review* 5(1), pp. 193–206.
34. Richard Thaler and Cass Sunstein (2008), *Nudge: Improving Decisions about Health, Wealth, and Happiness*, New Haven and London: Yale University Press.
35. For some early and profound insights, see Mancur Olson (1965/1971), *The Logic of Collective Action*, Cambridge, MA: Harvard University Press.
36. David J. Morrow (1999), 'Fen-Phen Maker to Pay Billions in Settlement of Diet-Injury Cases', *New York Times*, 8 October, and Bloomberg News reporting in the *New York Times* 'Wyeth in Settlement Talks Over Diet

Drugs', 19 January 2005. http://www.nytimes.com/1999/10/08/business/fen-phen-maker-to-pay-billions-in-settlement-of-diet-injury-cases.html (accessed 23 October 2017).

37. Bikhchandani, Hirshleifer and Welch (1992, 1998).
38. See Gerd Gigerenzer and Ulrich Hoffrage (1995), 'How to Improve Bayesian Reasoning Without Instruction: Frequency Formats', *Psychological Review* 102(4), pp. 684–704. For a survey of insights around the limits on Bayesian reasoning in humans, see Michelle Baddeley, Andrew Curtis and Rachel Wood (2004), 'An Introduction to Prior Information Derived from Probabilistic Judgments: Elicitation of Knowledge, Cognitive Bias and Herding', in *Geological Prior Information: Informing Science and Engineering*, ed. A. Curtis and R. Wood, Geological Society, London, Special Publications 239, pp. 15–27.

2 Mob psychology

1. George Klineman (1980), *The Cult That Died: The Tragedy of Jim Jones and the Peoples Temple*, New York: Putman Publishing Group; Tim Reiterman with John Jacobs (2008), *Raven: The Untold Story of the Rev. Jim Jones and His People*, New York: Penguin Books; https://en.wikipedia.org/wiki/Peoples_Temple (accessed 8 March 2017).
2. See Stanley Milgram (1963), 'Behavioral Study of Obedience', *Journal of Abnormal Psychology* 67, pp. 371–8; and Craig Haney, Curtis Banks and Philip Zimbardo (1973), 'A Study of Prisoners and Guards in a Simulated Prison', *Naval Research Review* 30, pp. 4–17.
3. For a comprehensive survey of this literature, see James Surowiecki (2004), *The Wisdom of Crowds: Why the Many Are Smarter Than the Few*, London: Abacus.
4. There is no complete English translation of the primary source: Marquis de Condorcet (1785), *Essay on the Application of Analysis to the Probability of Majority Decisions*. For a modern interpretation, see Krishna K. Ladha (1995), 'Information Pooling Through Majority-Rule Voting', *Journal of Economic Behavior and Organization* 26(3), pp. 353–72.
5. See https://www.ara.com/projects/aces-crowd-based-forecasting-world-events (accessed 23 October 2017).
6. David V. Budescu and Eva Chen (2015), 'Identifying Expertise to Extract the Wisdom of Crowds', *Management Science* 61(2), pp. 267–80.
7. While until recently emotions have been largely neglected in economics, Jon Elster has written extensively on how emotions can be brought into economic analyses; see for example Jon Elster (1996), 'Rationality and the Emotions', *Economic Journal* 106(438), pp. 136–97, and Jon Elster (1998), 'Emotions and Economic Theory', *Journal of Economic Literature* 36(1), pp. 47–74.
8. See Jaap van Ginneken (1992), *Crowds, Psychology, and Politics, 1871–1899*, Cambridge University Press.
9. Jean-Gabriel De Tarde (1890), *The Laws of Imitation*, trans. Elise Clews Parsons, 1903, New York: Henry Holt and Company.
10. Gustave Le Bon (1895), *The Crowd: A Study of the Popular Mind*, Lexington, KY: Maestro Reprints, p. 9.

11. Other cultures embrace the idea that the group is something quite different from the individuals who comprise it. The marketing expert Mark Earls explores some of these insights, including the Zulu and Xhosa concept of *Ubuntu* – which means something like 'shared humanity' – in Mark Earls (2009), *Herd: How to Change Mass Behaviour by Harnessing Our True Nature*, Chichester: John Wiley and Sons.
12. Le Bon (1895), pp. 11–12.
13. Charles Mackay (1841), *Extraordinary Popular Delusions and the Madness of Crowds*, Lexington, KY: Maestro Reprints.
14. For later psychoanalytic approaches, see the writings of other psychoanalysts, for example Ernest Jones (1923/2009), *Essays in Applied Psycho Analysis*, London: The International Psycho-Analytical Press, reprint by Charleston: BiblioBazaar.
15. Wilfred Trotter (1916), *Instincts of the Herd in Peace and War*, London: T. F. Unwin.
16. Sigmund Freud (1921), 'The Herd Instinct', in *Group Psychology and the Analysis of the Ego*, in *The Standard Edition of the Complete Psychological Works of Sigmund Freud*, Vol. XVIII, London: Vintage, pp. 117–21.
17. Trotter (1916), p. 120.
18. Aristotle, *Metaphysics*, Book VIII, 1045a.8–10.
19. Kurt Koffka (1935), *Principles of Gestalt Psychology*, New York: Harcourt, Brace and Company, p. 176.
20. Wilhelm Reich (1946), *The Mass Psychology of Fascism*, trans. Theodore P. Wolfe, New York: Orgone Institute Press.
21. Wilhelm Reich (1972), *Character Analysis*, 3rd edn, trans. Vincent R. Carfagno, ed. Mary Higgins and Chester M. Raphael, New York: Farrar, Straus and Giroux.
22. George A. Katona (1951), *Psychological Analysis of Economic Behavior*, New York: McGraw Hill; George A. Katona (1975), *Psychological Economics*, New York: Elsevier.
23. Akerlof and Kranton (2011).
24. Gustav Jahoda (2004), 'Henri Tajfel', *Oxford Dictionary of National Biography*, Oxford University Press; Stephen Reicher, 'Biography of Henri Tajfel (1919–1982)', European Association of Experimental Social Psychology. https://groups.google.com/forum/#!topic/jnu-psychology/zWK4S-ClSc0 (accessed 7 September 2017). 'Obituary: Henri Tajfel' (1982), *British Journal of Social Psychology* 21(3), pp. 185–8.
25. For a survey of the minimal group paradigm concept, see Michael Diehl (1990), 'The Minimal Group Paradigm: Theoretical Explanations and Empirical Findings', *European Review of Social Psychology* 1(1), pp. 263–92.
26. See for example Henri Tajfel (1970), 'Experiments in Intergroup Discrimination', *Scientific American* 223, pp. 96–102; Henri Tajfel, M.G. Billig, R.P. Bundy and Claude Flament (1971), 'Social Categorization and Intergroup Behaviour', *European Journal of Social Psychology* 1(2), pp. 149–78.
27. The economists Shaun Hargreaves Heap and Daniel Zizzo have explored some of these group influences from a group perspective using behavioural economics experiments; see Shaun P. Hargreaves Heap and Daniel John Zizzo (2009), 'The Value of Groups', *American Economic Review* 99(1), pp. 295–323.

28. Hipster mavericks are not new – the term 'hipster' was originally used in the context of 1940s jazz rebels in the US.
29. Ezra Klein (2015), 'On Paul Krugman's Theory of Hipsters', *Vox*, 27 July. http://www.vox.com/2015/7/27/9049025/paul-krugman-hipsters (accessed 7 September 2017).
30. Scientists have demonstrated that alcohol is associated with greater impulsivity and risk-taking; see for example Luca Corazzini, Antonio Filippin and Paolo Vanin (2015), 'Economic Behaviour under the Influence of Alcohol: An Experiment on Time Preferences, Risk-Taking, and Altruism', *PLoS ONE* 10(4). https://doi.org/10.1371/journal.pone.0121530 (accessed 7 September 2017).
31. Mark Levine, Robert Lowe, Rachel Best and Derek Heim (2012), '"We Police It Ourselves": Group Processes in the Escalation and Regulation of Violence in the Night-Time Economy', *European Journal of Social Psychology* 42, pp. 924–32.
32. Levine, Lowe, Best and Heim (2012), p. 927.
33. Levine, Lowe, Best and Heim (2012).
34. David Stout (1996), 'Obituary: Solomon Asch is Dead at 88; A Leading Social Psychologist', *New York Times*, 29 February. http://www.nytimes.com/1996/02/29/us/solomon-asch-is-dead-at-88-a-leading-social-psychologist.html (accessed 7 September 2017).
35. See for example Solomon Asch (1955), 'Opinions and Social Pressure', *Scientific American* 193(5), pp. 31–5. For a metaanalysis of experiments exploring the line judgement task, see Rod Bond and Peter B. Smith (1996), 'Culture and Conformity: A Meta-Analysis of the Studies Using Asch's (1952b, 1956) Line Judgment Task', *Psychological Bulletin* 119(1), pp. 111–37.
36. Asch's experiments capture something similar to experiments exploring conformity conducted by social psychologist Muzafer Sherif. He analysed social influences on people's perceptions of the 'autokinetic effect': when people are shown a light in a darkened room, they will mistakenly think that it is moving. Sherif found that, when he asked his experimental participants to announce estimates of how far the dots had travelled, the participants' estimates converged. See Muzafer Sherif (1935), *A Study of Some Social Factors in Perception*, New York: Archives of Psychology, No. 187. http://web.mit.edu/curhan/www/docs/Articles/15341_Readings/Influence_Compliance/Sherif_A_Study_of_Some_Social_Factors_(1935)_Arch%20Psych.pdf (accessed 30 October 2017).
37. Robert J. Shiller (1995), 'Conversation, Information and Herd Behavior', *American Economic Review* 85(2), pp. 181–5.
38. Solomon Asch (1952), *Social Psychology*, Englewood Cliffs, NJ: Prentice-Hall, p. 464.
39. A. Bandura, D. Ross and S.A. Ross (1961), 'Transmission of Aggression Through Imitation of Aggressive Models', *Journal of Abnormal and Social Psychology* 63, pp. 575–82.
40. There are a number of academic papers exploring these and similar findings; see for example Hunt Allcott (2011), 'Social Norms and Energy Conservation', *Journal of Public Economics* 95(9–10), pp. 1082–95; and Hunt Allcott and Todd Rogers (2014), 'The Short-Run and Long-Run Effects of Behavioral Interventions: Experimental Evidence from Energy Conservation', *American Economic Review* 104(10), pp. 3003–37.

41. These social 'nudges', as developed and promoted by the UK's Behavioural Insights Team, have been in the spotlight of recent policy debates and controversies; see for example Kate Palmer (2014), 'Psychology and "Nudges": Five Tricks the Taxman Uses to Make You Pay £210m Extra', *Daily Telegraph*, 9 October. http://www.telegraph.co.uk/finance/personal-finance/tax/11147321/Five-tricks-or-nudges-HMRC-uses-to-make-you-pay–210m-extra.html (accessed 7 September 2017); and Tamsin Rutter (2015), 'The Rise of Nudge – The Unit Helping Politicians to Fathom Human Behaviour', *Guardian*, 23 July. https://www.theguardian.com/public-leaders-network/2015/jul/23/rise-nudge-unit-politicians-human-behaviour (accessed 7 September 2017).
42. Subhrendu Pattanayak, Jui-Chen Yang, Katherine L. Dickinson, Christine Poulos, Sumeet R. Patil, Ranjan K. Mallick, Jonathan L. Blitstein and Purujit Praharaj (2009), 'Shame or Subsidy Revisited: Social Mobilization for Sanitation in Orissa, India', *Bulletin of the World Health Organization* 87, pp. 580–7.

3 Herding on the brain

1. Plato, *Phaedrus*, 246a–254e.
2. Colin F. Camerer, George Loewenstein and Drazen Prelec (2005), 'Neuroeconomics: How Neuroscience Can Inform Economics', *Journal of Economic Literature* 43(1), p. 9.
3. Daniel Kahneman (2011), *Thinking, Fast and Slow*, New York: Farrar, Straus and Giroux, develops earlier insights from modern neuroscience, as well as economics. He notes (p. 450) that the terms 'System 1' and 'System 2' are borrowed from Keith E. Stanovich and Richard F. West (2000), 'Individual Differences in Reasoning: Implications for the Rationality Debate', *Behavioral and Brain Sciences* 23, pp. 645–65. Other earlier work on similar themes includes Walter Schneider and Richard M. Shiffrin (1977), 'Controlled and Automatic Human Information Processing: Perceptual Learning, Automatic Attending and a General Theory', *Psychological Review* 84(2), pp. 127–90; and Paul M. Romer (2000), 'Thinking and Feeling', *American Economic Review* 90(2), pp. 439–43. For an earlier outline of Daniel Kahneman's systems model, see also Daniel Kahneman (2003), 'Maps of Bounded Rationality: Psychology for Behavioral Economics', *American Economic Review* 93(5), pp. 1449–75. For a simple introduction to dual process thinking, see Scott Barry Kaufman and Jerome L. Singer (2012), 'The Creativity of Dual Process "System 1" Thinking', *Scientific American* guest blog. https://blogs.scientificamerican.com/guest-blog/the-creativity-of-dual-process-system–1-thinking/ (accessed 7 September 2017).
4. Kahneman (2011), p. 25.
5. More controversially, some modern neuroscientists make a distinction between left-brain and right-brain dominance, with the left associated with more logical, mathematical styles of thinking, and the right associated with more emotional, creative styles of thinking. See Iain McGilchrist (2009), *The Master and His Emissary: The Divided Brain and the Making of the Modern World*, New Haven and London: Yale University Press.
6. Charles G. Gross (1998), 'Galen and the Squealing Pig', *History of Neuroscience* 4(3), pp. 216–21.

7. Gross (1998).
8. Le Bon (1895), pp. 13–14.
9. Stanley Finger (2001), *Origins of Neuroscience: A History of Explorations into Brain Function*, Oxford University Press.
10. Galen's surgery on pigs and their vocal cords was an example of ancient experimental evidence that our thinking is controlled from our brains. Cited in Gross (1998), p. 218.
11. John Martyn Harlow (1868), 'Recovery from the Passage of an Iron Bar Through the Head', *Publication of the Massachusetts Medical Society* 2, pp. 327–47; republished in M.B. Macmillan (2002), *An Odd Kind of Fame: Stories of Phineas Gage*, Cambridge, MA: MIT Press. See also Steve Twomey (2010), 'Phineas Gage: Neuroscience's Most Famous Patient', *Smithsonian Magazine*, January. https://www.smithsonianmag.com/history/phineas-gage-neurosciences-most-famous-patient-11390067/ (accessed 22 October 2017).
12. Antonio Damasio (1994/2006), *Descartes' Error: Emotion, Reason and the Human Brain*, London: Vintage.
13. Confidence in fMRI techniques is undermined by recent evidence that it is associated with too many false positives. In other words, there are lots of statistically significant findings which may be spurious, simply a result of the techniques. For example, Anders Eklund and colleagues found a false positive rate of 70 per cent in their fMRI analyses; see Anders Eklund, Thomas E. Nichols and Hans Knutsson (2016), 'Cluster Failure: Why fMRI Inferences for Spatial Extent Have Inflated False-Positive Rates', *Proceedings of the National Academy of Sciences of the United States of America* 113(28), pp. 7900–5.
14. Wim De Neys, Oshin Vartanian and Vinod Goel (2008), 'Smarter Than We Think: When Our Brains Detect That We Are Biased', *Psychological Science* 19(5), pp. 483–9.
15. For the original Engineer-Lawyer study, see Daniel Kahneman and Amos Tversky (1973), 'On the Psychology of Prediction', *Psychological Review* 80, pp. 237–51.
16. For a layperson's introduction to some neuroscientific evidence about social instincts, explored in the context of a range of modern issues and problems, see Peter Bazalgette (2017), *The Empathy Instinct: How to Create a More Civil Society*, London: John Murray.
17. The pioneering neuroeconomist Colin Camerer and colleagues have outlined the key ideas underlying neuroeconomics in useful survey articles; see for example Colin F. Camerer, George Loewenstein and Drazen Prelec (2004), 'Neuroeconomics: Why Economics Needs Brains', *Scandinavian Journal of Economics* 106(3), pp. 555–79; and Camerer, Loewenstein and Prelec (2005). See also Michelle Baddeley (2013), *Behavioural Economics and Finance*, Abingdon: Routledge, ch. 3: 'Foundations: Neuroscience and Neuroeconomics', pp. 30–47.
18. See Wolfram Schultz, Peter Dayan and P. Read Montague (1997), 'A Neural Substrate of Prediction and Reward', *Science* 275(5306), pp. 1593–9; and Wolfram Schultz (2002), 'Getting Formal with Dopamine and Reward', *Neuron* 36, pp. 241–63.
19. For a simple introduction to the concepts see Wolfram Schultz (2007), 'Reward Signals', *Scholarpedia* 2(6), p. 2184. http://www.scholarpedia.org/article/Reward_signals (accessed 30 October 2017).

20. Other similar studies include a series of imaging experiments led by Vasily Klucharev, a neuroscientist based at the Donders Centre for Cognitive Neuroimaging in the Netherlands. With his colleagues, Klucharev has connected the idea of conformity with reinforcement learning, linking with some of Shultz's ideas about reward prediction error. See Vasily Klucharev, Kaisa Hytönen, Mark Rijpkema, Ale Smidts and Guillén Fernández (2009), 'Reinforcement Learning Signal Predicts Social Conformity', *Neuron* 61(1), pp. 140–51. For an introduction to some of the key connections between the economic and neuroscientific literature about social emotions driving herding, see Michelle Baddeley (2010), 'Herding, Social Influence and Economic Decision-Making: Socio-Psychological and Neuroscientific Analyses', *Philosophical Transactions of the Royal Society B* 365(1538), pp. 281–90.
21. The images of the faces were generated from a face bank. The faces did not have hair – different hairstyles can apparently distort our assessment of the reliability of a person's judgement. We incorporated an additional control scenario in which people were shown faces of monkeys instead of humans, to capture if the participants were responding to any type of face, regardless of whether they could assume that the faces knew anything about share-trading.
22. Christopher Burke, Michelle Baddeley, Philippe Tobler and Wolfram Schultz (2010), 'Striatal BOLD Response Reflects the Impact of Herd Information on Financial Decisions', *Frontiers in Human Neuroscience* 4, article 48. https://doi.org/10.3389/fnhum.2010.00048 (accessed 5 September 2017).
23. See Alan G. Sanfey, James K. Rilling, Jessica A. Aronson, Leigh E. Nystrom and Jonathan D. Cohen (2003), 'The Neural Basis of Economic Decision-Making in the Ultimatum Game', *Science* 300, pp. 1755–8. A similar neuroeconomic study of brain activations in the context of decisions that catalyse conflicts between the emotional and the cognitive was conducted by Samuel M. McClure, David I. Laibson, George Loewenstein and John D. Cohen (2004), 'Separate Neural Systems Value Immediate and Delayed Rewards', *Science* 313, pp. 684–7: they studied similar influences in the context of decisions over time, when there is a struggle between our patient and impatient selves.
24. The ultimatum game was first implemented by Werner Güth, Rolf Schmittberger and Bernd Schwarze (1982), 'An Experimental Analysis of Ultimatum Bargaining', *Journal of Economic Behavior and Organization* 3(4) pp. 367–88. For a cross-cultural study, see Joseph Henrich, Robert Boyd, Samuel Bowles, Colin Camerer, Ernst Fehr, Herbert Gintis and Richard McElreath (2001), 'In Search of *Homo economicus*: Behavioural Experiments in 15 Small-Scale Societies', *American Economic Review* 91(2), pp. 73–8. See also Joseph Henrich, Robert Boyd, Samuel Bowles, Colin Camerer, Ernst Fehr, Herbert Gintis and Richard McElreath (2004), *Foundations of Human Sociality: Economic Experiments and Ethnographic Evidence from Fifteen Small-Scale Societies*, Oxford University Press. For a short summary of the ultimatum game and other games used by behavioural economists to pick up pro-social preferences, see Michelle Baddeley (2017), *Behavioural Economics: A Very Short Introduction*, Oxford University Press, ch. 3: 'Social Lives', pp. 19–33.

25. Ramsey M. Raafat, Nick Chater and Chris Frith (2009), 'Herding in Humans', *Trends in Cognitive Sciences* 13(10), pp. 420–8.
26. Herbert Simon is attributed with the first analyses of heuristics – for example, see Herbert A. Simon and Allen Newell, 'Heuristic Problem Solving: The Next Advance in Operations Research', *Operations Research* 6(1), pp. 1–10. For early mentions of herding heuristics, see Michelle Baddeley, Demetris Pillas, Yorgos Christopoulos, Wolfram Schultz and Philippe Tobler (2007), 'Herding and Social Pressure in Trading Tasks: A Behavioural Analysis', Cambridge Working Papers in Economics 0730, Faculty of Economics, University of Cambridge. https://doi.org/10.17863/CAM.5145 (accessed 7 September 2017); Baddeley, Curtis and Wood (2004).
27. See, for example, Gerd Gigerenzer and Daniel Goldstein (1996), 'Reasoning the Fast and Frugal Way: Models of Bounded Rationality', *Psychological Review* 103, pp. 650–9.

4 Animal herds

1. Rory Tingle (2016), 'The Power of the Herd: Thousands of Wildebeest Make It Across a Crocodile-Infested River by Stampeding Together in a Huge Crowd', *Mail Online*, 2 August. http://www.dailymail.co.uk/news/article-3720191/The-power-herd-Thousands-wildebeest-make-crocodile-infested-river-migration-Kenya-Tanzania-stampeding-huge-crowd.html (accessed 7 September 2017).
2. C. Muro, R. Escobedo, L. Spector and R.P. Coppinger, 'Wolf-Pack (*Canis lupus*) Hunting Strategies Emerge from Simple Rules in Computational Simulations', *Behavioural Processes* 88(3), pp. 192–7.
3. For a collection of articles from social neuroscience exploring social learning and related themes, see Steve W.C. Chang and Masaki Isoda (2015), *Neural Basis of Social Learning, Social Deciding, and Other-Regarding Preferences*, Lausanne: Frontiers in Psychology. https://www.frontiersin.org/books/Neural_basis_of_social_learning_social_deciding_and_other-regarding_preferences/449 (accessed 20 October 2017).
4. Étienne Danchin, L.-S. Giraldeau, T.J. Valone and R.H. Wagner (2004), 'Public Information: From Nosy Neighbours to Cultural Evolution', *Science* 305, pp. 487–91.
5. A. Kis, L. Huber and A. Wilkinson (2015), 'Social Learning by Imitation in a Reptile (*Pogona vitticeps*)', *Animal Cognition* 18(1), pp. 325–31.
6. F. Cortesi, W.E. Feeney, J. Marshall and K.L. Cheney (2015), 'Phenotypic Plasticity Confers Multiple Fitness Benefits to a Mimic', *Current Biology* 25, pp. 1–6.
7. Nicholas B. Davies, John R. Krebs and Stuart A. West (2012), *An Introduction to Behavioural Ecology*, 4th edn, Oxford: Wiley Blackwell, p. 148.
8. H. Kruuk (1964), 'Predators and Anti-Predator Behaviour of the Black-Headed Gull *Larus ridibundus*', *Behaviour* 11, pp. 1–129; D.B.A. Thompson and C.J. Barnard (1983), 'Anti-Predator Responses in Mixed-Species Associations of Lapwings, Golden Plovers and Black-Headed Gulls', *Animal Behaviour* 31(2), pp. 585–93.
9. K.M. McLennan (2012), 'Farmyard Friends', *Biologist* 59(4), pp. 18–22.
10. Jonathan D. Cohen (2005), 'The Vulcanization of the Human Brain: A Neural Perspective on Interactions between Cognition and Emotion', *Journal of Economic Perspectives* 19(4), pp. 3–24.

11. Danchin, Giraldeau, Valone and Wagner (2004).
12. Corina E. Tarnita, Alex Washburne, Ricardo Martinez-Garcia, Allyson E. Sgro and Simon A. Levin (2015), 'Fitness Trade-Offs Between Spores and Nonaggregating Cells Can Explain the Coexistence of Diverse Genotypes in Cellular Slime Molds', *Proceedings of the National Academy of Sciences* 112(9), pp. 2776–81.
13. Paul B. Rainey (2015), 'Precarious Development: The Uncertain Social Life of Cellular Slime Molds', *Proceedings of the National Academy of Sciences of the United States of America* 112(9), pp. 2639–40.
14. See Alan Kirman (1993), 'Ants, Rationality and Recruitment', *Quarterly Journal of Economics* 108(1), pp. 137–56.
15. Edward O. Wilson and Bert Hölldobler (2005), 'Eusociality: Origin and Consequences', *Proceedings of the National Academy of Sciences of the United States of America* 102(38), pp. 13367–71.
16. Davies, Krebs and West (2012).
17. Le Bon (1895), p. 11.
18. Tingle (2016).
19. For example see Kristen Hawkes, Jonathan F. O'Connell, Nicholas G. Blurton Jones, Helen Alvarez and Eric L. Charnov, 'Grandmothering, Menopause, and the Evolution of Human Life Histories', *Proceedings of the National Academy of Sciences of the United States of America* 95(3), pp. 1335–9.
20. Davies, Krebs and West (2012), pp. 77–8.
21. Colin Camerer (2003), 'Strategizing in the Brain', *Science* 300, pp. 1673–5.
22. Richard Dawkins (1976), *The Selfish Gene*, Oxford University Press. See also Susan Blackmore (1999), *The Meme Machine*, Oxford University Press; and Aaron Lynch (1996), *Thought Contagion: How Belief Spreads Through Society*, New York: Basic Books.
23. Scientists including E.O. Wilson and Stephen Jay Gould, as well as the economist John Maynard Smith, focus on the role of organisms in evolution, arguing against Dawkins' focus on genes as the vehicles for natural selection. For further reading on these debates, including economists' insights, see also Robert Axelrod (1984), *The Evolution of Cooperation*, Cambridge, MA: Basic Books; Stephen Jay Gould (1992), *The Panda's Thumb: More Reflections in Natural History*, New York: W.W. Norton; Stephen Jay Gould (2001), *The Lying Stones of Marrakech: Penultimate Reflections in Natural History*, London: Vintage; John Maynard Smith (1982), *Evolution and the Theory of Games*, Cambridge University Press; and John Maynard Smith (1974), 'The Theory of Games and the Evolution of Animal Conflicts', *Journal of Theoretical Biology* 47, pp. 209–21.
24. For some fascinating surveys of culture and other social behaviours in animals, see Hal Whitehead and Luke Rendell (2015), *The Cultural Lives of Whales and Dolphins*, University of Chicago Press; Carl Safina (2015), *Beyond Words: What Animals Think and Feel*, New York: John Macrae/Henry Holt and Company; and Kieran C.R. Fox, Michael Muthukrishna and Susanne Shultz (2017), 'The Social and Cultural Roots of Whale and Dolphin Brains', *Nature Ecology and Evolution*. https://www.nature.com/articles/s41559-017-0336-y (accessed 30 September 2017).
25. Davies, Krebs and West (2012), pp. 76–7.

26. Richard W. Wrangham, W.C. McGrew, Frans B.M. de Waal and Paul G. Heltne (eds) (1996), *Chimpanzee Cultures*, Cambridge, MA: Harvard University Press; Michael Balter (2013), 'Strongest Evidence of Animal Culture Seen in Monkeys and Whales', *Science*, 25 April. http://www.sciencemag.org/news/2013/04/strongest-evidence-animal-culture-seen-monkeys-and-whales (accessed 7 September 2017).
27. Luke Rendell and Hal Whitehead (2001), 'Culture in Whales and Dolphins', *Behavioral and Brain Sciences* 24, pp. 309–82.
28. Gene S. Helfman and Eric T. Shultz (1984), 'Social Transmission of Behavioural Traditions in a Coral Reef Fish', *Animal Behaviour* 32, pp. 379–84.
29. This idea is known as the 'social brain hypothesis'. For a survey of the idea see Robin I.M. Dunbar (2009), 'The Social Brain Hypothesis and Its Implications for Social Evolution', *Annals of Human Biology* 36(5), pp. 562–72.
30. Harry J. Jerison (1973), *Evolution of the Brain and Intelligence*, New York and London: Academic Press; Paul D. MacLean (1990), *The Triune Brain in Evolution: Role in Paleocerebral Functions*, New York: Springer.
31. For an analysis of human cooperation from an evolutionary perspective, see David G. Rand and Martin A. Novak (2013), 'Human Cooperation', *Trends in Cognitive Sciences* 17(8), pp. 413–25.
32. Herbert Simon (1990), 'A Mechanism for Social Selection and Successful Altruism', *Science* 250, pp. 1665–8.
33. Tania Singer, Ben Seymour, John O'Doherty, Holger Kaube, Raymond J. Dolan and Chris D. Frith (2004), 'Empathy for Pain Involves the Affective But Not Sensory Components of Pain', *Science* 303(5661), pp. 1157–62. See also Patricia L. Lockwood (2016), 'The Anatomy of Empathy: Vicarious Experience and Disorders of Social Cognition', *Behavioural Brain Research* 311, pp. 255–66.
34. Tania Singer and Ernst Fehr (2005), 'The Neuroeconomics of Mind-Reading and Empathy', *American Economic Review* 95(2), pp. 340–5.
35. Una Frith and Chris Frith (2003), 'Development and Neurophysiology of Mentalizing', *Philosophical Transactions of the Royal Society B* 358(1431), pp. 459–73; Chris D. Frith and Una Frith (2006), 'The Neural Basis of Mentalizing', *Neuron* 50(4), pp. 531–4.
36. Elisabeth Hill and David F. Sally (2003), 'Dilemmas and Bargains: Autism, Theory-of-Mind, Cooperation and Fairness'. https://ssrn.com/abstract=407040 (accessed 25 October 2017).
37. See also Alessio Aventani, Domenica Bueti, Gaspare Galati and Salvatore M. Aglioti (2005), 'Transcranial Magnetic Stimulation Highlights the Sensorimotor Side of Empathy for Pain', *Nature Neuroscience* 8(7), pp. 955–60 for transcranial magnetic stimulation (TMS) studies of empathising with others' pain.
38. Franco Cauda, Giuliano Carlo Geminiani and Alessandro Vercelli (2014), 'Evolutionary Appearance of Von Economo's Neurons in the Mammalian Cerebral Cortex', *Frontiers in Human Neuroscience* 8, article 104.
39. See Giacomo Rizzolatti and Laila Craighero (2004), 'The Mirror Neuron System', *Annual Review of Neuroscience* 27(1), pp. 169–92; Giacomo Rizzolatti (2005), 'The Mirror Neuron-System and Imitation', in *Perspectives on Imitation: From Neuroscience to Social Science*, vol. 1,

Mechanisms of Imitation and Imitation in Animals, ed. Susan Hurley and Nick Chater, Cambridge, MA: MIT Press, pp. 55–76; Marco Iacoboni (2009), 'Imitation, Empathy, and Mirror Neurons', *Annual Review of Psychology* 60, pp. 653–70; and Marco Iacoboni, Roger P. Woods, Marcel Brass, Harold Bekkering, John C. Mazziotta and Giacomo Rizzolatti (1999), 'Cortical Mechanisms of Human Imitation', *Science* 286, pp. 2526–8. Baddeley, Curtis and Wood (2004) suggest that this mirror neuron activity may explain imitation and herding in socioeconomic contexts.

40. Rizzolatti and Craighero (2004).
41. For some general analyses of the evolution of human behaviour, see Paul Seabright (2004), *The Company of Strangers: A Natural History of Economic Life*, Princeton University Press. For a more historical perspective, see Yuval Noah Harari (2014), *Sapiens: A Brief History of Humankind*, London: Harvill Secker/Random House. See also '*Homo sapiens*', Smithsonian National Museum of Natural History online, 29 June 2017. http://humanorigins.si.edu/evidence/human-fossils/species/homo-sapiens (accessed 7 September 2017).
42. See, for example, Elaine Hatfield, John T. Cacioppo and Richard L. Rapson (1994), *Emotional Contagion*, Cambridge University Press.
43. Raafat, Chater and Frith (2009).
44. Cohen (2005).

5 Mavericks

1. Christopher Turner (2011), 'Wilhelm Reich: The Man Who Invented Free Love', *Guardian*, 8 July. https://www.theguardian.com/books/2011/jul/08/wilhelm-reich-free-love-orgasmatron (accessed 7 September 2017).
2. Upasana Chauhan (2016), 'Why I Risked an Honor Killing to Reject an Arranged Marriage', *TIME*. http://time.com/4450544/arranged-marriage-india/ (accessed 30 September 2017).
3. For a survey on the literatures covering the negative impacts, see Kipling D. Williams, Joseph P. Forgas, Willian von Hippel and Lisa Zadro (eds) (2005), *The Social Outcast: Ostracism, Social Exclusion, Rejection, and Bullying*, The Sydney Symposium of Social Psychology series, New York/Hove: Psychology Press/Taylor & Francis Group.
4. Keynes (1936), p. xxiii.
5. W. Heath Robinson (2007), *Contraptions*, London: Duckworth; 'William Heath Robinson', Heath Robinson Museum online. https://www.heathrobinsonmuseum.org/williamheathrobinson (accessed 7 September 2017).
6. The William Heath Robinson Museum in London is a monument to his whimsical contributions. See Olivia Solon (2016), 'William Heath Robinson Museum Finally Opens This Weekend. Who is the Man Behind the Legend?', *Wired*, 13 October. http://www.wired.co.uk/article/heath-robinson-deserves-a-museum (accessed 7 September 2017).
7. David Hirshleifer and Robert Noah (1998), 'Misfits and Social Progress', in Robert Noah (1998), *Essays in Learning and the Revelation of Private Information*, PhD thesis, University of Michigan, ProQuest Dissertations Publishing.
8. Hirshleifer and Noah (1998).

9. See for example Chunpeng Yang (2011), 'Maverick Risk of Individuals' Risk Perceptions', *Journal of Systems Science and Information* 9(2), pp. 119–26. We will explore the issue of maverick risk in more detail in chapter 6.
10. Bernheim (1994).
11. The seminal text for expected utility theory is John von Neumann and Oskar Morgenstern (1944), *Theory of Games and Economic Behavior*, Princeton University Press.
12. 'Sticky Lawsuit: $400M Dispute Lingers Over Post-It Inventor', *Chicago Tribune*, 11 March 2016. http://www.chicagotribune.com/business/ct-post-it-note-inventor-lawsuit-20160311-story.html (accessed 7 September 2017).
13. Daniel Kahneman and Amos Tversky (1979), 'Prospect Theory: An Analysis of Decision under Risk', *Econometrica* 47(2), pp. 263–92.
14. See Amos Tversky and Daniel Kahneman (1974), 'Judgement under Uncertainty: Heuristics and Bias', *Science* 185(4157), pp. 1124–31. Also see the collected volumes Daniel Kahneman, Paul Slovic and Amos Tversky (eds) (1982), *Judgement under Uncertainty: Heuristics and Biases*, Cambridge University Press, and Daniel Kahneman and Amos Tversky (eds) (2000), *Choices, Values and Frames*, Cambridge University Press.
15. Kahneman and Tversky (1979).
16. For a technical analysis of these different hypotheses in the context of herding and contrarianism in financial markets, see Chad Kendall (2017), 'Herding and Contrarianism: A Matter of Preference', paper presented to the Financial Management Association Conference, Boston, MA, July. http://www.fmaconferences.org/Boston/kendall-ptherding.pdf (accessed 14 September 2017).
17. John Beshears, James J. Choi, David Laibson, Brigitte C. Madrian and Katherine L. Milkman (2015), 'The Effect of Providing Peer Information on Retirement Savings Decisions', *Journal of Finance* LXX(3), pp. 1161–201.
18. The *Harvard Business Review* has a range of articles covering the various aspects of the entrepreneurial personality: for example, see Timothy Butler (2017), 'Hiring an Entrepreneurial Leader', *Harvard Business Review*, March–April, pp. 84–93, and Daniel McGinn (2016), 'Is There a Connection Between Entrepreneurship and Mental Health Conditions?', *Harvard Business Review*, 22 February. https://hbr.org/2016/02/222-entrepreneurship-ic-do-narcissists-make-great-entrepreneurs (accessed 7 September 2017).
19. Karen Blumenthal (2012), *Steve Jobs: The Man Who Thought Different*, London: Bloomsbury.
20. Kahneman (2011).
21. For more on the idea that risk is a feeling, see George F. Loewenstein, Elke U. Weber, Christopher K. Hsee and Ned Welch (2001), 'Risk as Feelings', *Psychological Bulletin* 127(2), pp. 267–86; and Paul Slovic (2010), *The Feeling of Risk: New Perspectives on Risk Perception*, London: Earthscan.
22. Burke, Baddeley, Tobler and Schultz (2010), pp. 8–9.
23. Cass R. Sunstein (2002), 'Conformity and Dissent', John M. Olin Law and Economics Working Paper No. 164, The Law School, University of Chicago. http://chicagounbound.uchicago.edu/public_law_and_legal_theory/68/ (accessed 5 September 2017).

24. For example, see David S. Landes, Joel Mokyr and William J. Baumol (eds) (2012), *The Invention of Enterprise: Entrepreneurship from Ancient Mesopotamia to Modern Times*, Princeton University Press.
25. Gavin Weightman (2015), *Eureka: How Invention Happens*, New Haven and London: Yale University Press.
26. On a similar track, economists Andrew Clark and Andrew Oswald develop the idea that individuals who want to be different may rationally decide to imitate: see Andrew E. Clark and Andrew J. Oswald (1998), 'Comparison-Concave Utility and Following Behaviour in Social and Economic Settings', *Journal of Public Economics* 70, pp. 133–55.
27. Quoted in Andrew Sinclair (2006), *Viva Che! The Strange Death and Life of Che Guevara*, Stroud: Sutton Publishing.
28. Quoted in Sinclair (2006).
29. For a contrasting view of fads and fashions, see Rolf Meyerson and Elihu Katz (1957), 'Notes on a Natural History of Fads', *American Journal of Sociology* 62(6), pp. 594–601.
30. Peter Beaumont, Antony Barnett and Gaby Hinsliff (2003), 'Iraqi Mobile Labs Nothing to do with Germ Warfare, Report Finds', *Guardian*, 5 June. https://www.theguardian.com/world/2003/jun/15/Iraq (accessed 7 September 2017); Justin Ling (2016), 'The UK's Iraq War Inquiry Vindicates a Whistleblower Who Took His Own Life', *Vice*, 6 July. https://news.vice.com/article/the-uks-iraq-war-inquiry-vindicates-david-kelly-a-whistleblower-who-took-his-own-life (accessed 7 September 2017).
31. Kamal Ahmed (2003), 'Revealed: How Kelly Article Set Out Case for War in Iraq', *Guardian*, 31 August. https://www.theguardian.com/politics/2003/aug/31/davidkelly.iraq1 (accessed 7 September 2017).
32. This verdict was controversial, and some doctors publicly expressed their scepticism in a letter to the *Guardian*; see David Halpin, C. Stephen Frost and Searle Sennett (2004), 'Our Doubts about Dr Kelly's Suicide', *Guardian*, 27 January. https://www.theguardian.com/theguardian/2004/jan/27/guardianletters4 (accessed 7 September 2017).
33. For example, Whistle-blowers Australia, an association instituted to protect the rights of those exposing corruption, malpractice and other wrongdoing: www.whistleblowers.org.au (accessed 7 September 2017). The US has introduced a range of legislation to protect whistleblowers, though the reach of this legislation is limited, especially those provisions relating to intelligence and national security: 'Whistleblower Protection in the United States', Wikipedia. https://en.wikipedia.org/wiki/Whistleblower_protection_in_the_United_States (accessed 7 September 2017). The UK government, meanwhile, provides advice for whistle-blowers: 'Whistleblowing for Employees: What is a Whistleblower?' https://www.gov.uk/whistleblowing/what-is-a-whistleblower (accessed 7 September 2017). For a list of modern whistleblowers see 'List of Whistleblowers: 2000s', Wikipedia. https://en.wikipedia.org/wiki/List_of_whistleblowers#2000s (accessed 7 September 2017).

6 Entrepreneurs versus speculators

1. There are numerous biographies, of which Skidelsky's is the most informative and engaging; see Robert Skidelsky (2005), *John Maynard Keynes: 1883–1946 – Economist, Philosopher, Statesman*, London: Penguin Books.

2. David Chambers, Elroy Dimson and Justin Food (2015), 'Keynes the Stock Market Investor: A Quantitative Analysis', *Journal of Financial and Quantitative Analysis* 50(4), pp. 431–49.
3. 'Keynesian Investment: Returns Fit for King's', *The Economist*, 22 June 2012. http://www.economist.com/blogs/freeexchange/2012/06/keynesian-investment (accessed 7 September 2017); David Chambers, Elroy Dimson and Justin Foo (2013), 'Keynes the Stock Market Investor: A Quantitative Analysis', *Journal of Financial and Quantitative Analysis* 50(4), pp. 431–49.
4. For a history of financial crises see Charles P. Kindleberger and Robin Aliber (2005), *Manias, Panics and Crashes: A History of Financial Crises*, 5th edn, Hoboken, NJ: John Wiley and Sons.
5. The conventional use of money relates to the idea that money is a social institution, and this insight is cogently and comprehensively explored by Geoffrey Ingham (2013), *The Nature of Money*, Cambridge: Polity Press.
6. For an analysis of the economic implications of e-cash see Michelle Baddeley (2004), 'Using e-Cash in the New Economy: An Economic Analysis of Micropayments Systems', *Journal of Electronic Commerce Research* 5(4), pp. 239–53.
7. See https://bristolpound.org and https://brixtonpound.org (accessed 7 September 2017).
8. There is an extensive range of economic analysis of speculative herding, including some lab experiments. See for example M. Cipriani and A. Guarino (2005), 'Herd Behavior in a Laboratory Financial Market', *American Economic Review* 95(5), pp. 1427–43, and Mathias Drehmann, Jörg Oechssler and Andreas Roider (2005), 'Herding and Contrarian Behavior in Financial Markets: An Internet Experiment', *American Economic Review* 95(5), pp. 1403–26.
9. For an economic analysis of some famous examples, see Peter M. Garber (2001), *Famous First Bubbles: The Fundamentals of Early Manias*, Cambridge, MA: MIT Press. For an extensive range of other examples of speculative manias, see Mackay (1841).
10. See Edward Chancellor (1998), *Devil Take the Hindmost: A History of Financial Speculation*, New York/London: Plume/Penguin Books.
11. Generally, economists struggle to get an objective handle on value, one of the critiques highlighted in the Cambridge capital controversy of the 1970s; see for example Geoffrey C. Harcourt (1972), *Some Cambridge Controversies in the Theory of Capital*, Cambridge University Press.
12. The original formulation of the rational expectations hypothesis is attributed to John Fraser Muth (1961), 'Rational Expectations and the Theory of Price Movements', *Econometrica* 29(3), pp. 315–35.
13. For a review, see Eugene Fama (1970), 'Efficient Capital Markets: A Review of Theory and Empirical Work', *Journal of Finance* 25, pp. 383–417.
14. Keynes (1936), ch. 12. See also John Maynard Keynes (1937), 'The General Theory of Employment', *Quarterly Journal of Economics* 51(2), pp. 209–23.
15. Keynes (1936, 1937).
16. Friedrich Hayek, a pioneer from the Austrian School of economics, proposed an alternative view to Keynes in exploring how knowledge unfolds as the product of social interactions. According to Hayek, as people observe a situation one by one, information is processed serially by

each successive decision maker, creating a path-dependent process: other people's past beliefs dictate the beliefs of their successors. See Friedrich Hayek (1952), *The Sensory Order: An Inquiry into the Foundations of Theoretical Psychology*, University of Chicago Press; Salvatore Rizzello (2004), 'Knowledge as a Path-Dependence Process', *Journal of Bioeconomics* 6(3), pp. 255–74.

17. There is continuity in the development of Keynes' ideas about probabilistic judgements – as outlined in John Maynard Keynes (1921), *A Treatise on Probability*, London: Macmillan and Co., and his development of insights around conventions and social influences in the macroeconomy and financial system, as outlined in Keynes (1936, 1937).
18. Itzhak Venezia, Amrut Nashikkar and Zur Shapira (2011), 'Firm Specific and Macro Herding by Professional and Amateur Investors and Their Effects on Market Volatility', *Journal of Banking and Finance* 35, pp. 1599–609.
19. Richard Topol (1991), 'Bubbles and Volatility of Stock Prices: Effect of Mimetic Contagion', *Economic Journal* 101(407), pp. 786–800. There is an extensive economics literature on financial herding: for example, Andrea Devenow and Ivo Welch (1996), 'Rational Herding in Financial Economics', *European Economic Review* 40, pp. 603–15, and Christopher Avery and Peter Zemsky (1998), 'Multidimensional Uncertainty and Herd Behavior in Financial Markets', *American Economic Review* 88(4), pp. 724–48.
20. D.S. Scharfstein and J.C. Stein (1990), 'Herd Behavior and Investment', *American Economic Review* 80(3), pp. 465–79.
21. Scharfstein and Stein (1990). For an analysis of group influences in the asset management market, see also Anna Tilba, Michelle Baddeley and Yixi Liao (2016), 'The Effectiveness of Oversight Committees: Decision-Making, Governance, Costs and Charges', Financial Conduct Authority Asset Management Market Study interim report. https://www.fca.org.uk/publication/research/tilba-baddeley-liao.pdf (accessed 7 September 2017).
22. Keynes (1936).
23. Colin F. Camerer (2003), *Behavioral Game Theory: Experiments in Strategic Interaction*, New York/Princeton, NJ: Russell Sage Foundation/Princeton University Press, pp. 216–17.
24. Keynes (1936), p. 59.
25. Baddeley (2010).
26. Mark J. Kamstra, Lisa A. Kramer and Maurice Levi (2003), 'Winter Blues: A SAD Stock Market Cycle', *American Economic Review* 93(1), pp. 324–43.
27. David Hirshleifer and Tyler Shumway (2003), 'Good Day Sunshine: Stock Returns and the Weather', *Journal of Finance* 58(3), pp. 1009–32.
28. Robert Prechter (2016), *The Socionomic Theory of Finance*, Gainesville, GA: Socionomics Institute Press. See also John L. Casti (2010), *Mood Matters: From Rising Skirt Lengths to the Collapse of World Powers*, Berlin: Springer-Verlag.
29. An analysis of links between Tulipmania, speculation and economic/financial theory can be found in Michelle Baddeley and John McCombie (2001/2004), 'An Historical Perspective on Speculative Bubbles and Financial Crisis: Tulipmania and the South Sea Bubble', in *What Global*

Economic Crisis?, ed. P. Arestis, M. Baddeley and J. McCombie, London: Palgrave Macmillan, and Michelle Baddeley (2018), 'Financial Instability and Speculative Bubbles: Behavioural Insights and Policy Implications', in *Alternative Approaches in Macroeconomics: Essays in Honour of John McCombie*, ed. Philip Arestis, London: Palgrave Macmillan, pp. 209–34.

30. Keynes (1936), p. 163.
31. Andrew W. Lo, Dmitry V. Repin and Brett N. Steenbarger (2005), 'Fear and Greed in Financial Markets: A Clinical Study of Day Traders', *American Economic Review* 95(2), pp. 352–9.
32. Shlomo Benartzi and Richard H. Thaler (1995), 'Myopic Loss Aversion and the Equity Premium Puzzle', *Quarterly Journal of Economics* 110(1), pp. 73–92.
33. See Hyman P. Minsky (1986), *Stabilising an Unstable Economy*, New Haven and London: Yale University Press; Hyman P. Minsky (1992), 'The Financial Instability Hypothesis', Levy Economics Institute Working Paper No. 74, Annandale on Hudson, NY: The Jerome Levy Economics Institute of Bard College. http://www.levy.org/pubs/wp74.pdf (accessed 5 March 2018).
34. Loewenstein, Weber, Hsee and Welch (2001); reprinted in G.F. Loewenstein (ed.) (2007), *Exotic Preferences: Behavioral Economics and Human Motivation*, Oxford University Press, pp. 567–611.
35. Joseph A. Schumpeter (1934/1981), *The Theory of Economic Development: An Inquiry into Profits, Capital, Credit, Interest, and the Business Cycle*, trans. John E. Elliott, New Brunswick, NJ: Transaction Books.
36. Keynes (1936), pp. 157–8.
37. Daron Acemoğlu (1992), 'Learning about Others' Actions and the Investment Accelerator', *Economic Journal* 103(417), pp. 318–28.
38. Francisco Campos, Michael Frese, Markus Goldstein, Leonardo Iacovone, Hillary C. Johnson, David McKenzie and Mona Mensmann (2017), 'Teaching Personal Initiative Beats Traditional Training in Boosting Small Business in West Africa', *Science* 357, pp. 1287–90. See also 'Teaching Entrepreneurship: Mind Over Matter', *The Economist*, 23 September 2017, p. 69.
39. Keynes (1936), p. 161.
40. Jerome Kagan (1998), *Galen's Prophecy: Temperament in Human Nature*, New York: Basic Books.
41. Keynes (1936), p. 150.
42. George Akerlof and Robert Shiller (2009), *Animal Spirits: How Human Psychology Drives the Economy and Why It Matters for Global Capitalism*, Princeton University Press. In terms of this book's definition of animal spirits, many have argued that Akerlof and Shiller's account is not true to Keynes' original vision of entrepreneurial personalities and motivations. See also Michelle Baddeley (2009), 'Far from a Rational Crowd: review of G. Akerlof and R. Shiller (2010), "Animal Spirits: How Human Psychology Drives the Economy"', *Science* 324, pp. 883–4.
43. Akerlof and Shiller (2009), pp. 153–6.
44. For a comprehensive account of who was responsible for the 2007/08 financial crisis, see Howard Davies (2010), *The Financial Crisis: Who Is to Blame?* Cambridge: Polity Press.
45. Minsky (1986, 1992).

46. Paul Ormerod (1998), *Butterfly Economics: A New General Theory of Economic and Social Behaviour*, London: Faber and Faber.
47. Tilba, Baddeley and Liao (2016).

7 Herding experts

1. Debbie Cenziper (2015), 'A Disputed Diagnosis Imprisons Parents', *Washington Post*, 20 March. https://www.washingtonpost.com/graphics/investigations/shaken-baby-syndrome/ (accessed 27 October 2017).
2. John Caffey (1974), 'The Whiplash Shaken Infant Syndrome: Manual Shaking by the Extremities with Whiplash-Induced Intracranial and Intraocular Bleedings, Linked with Residual Permanent Brain Damage and Mental Retardation', *Pediatrics* 54(4), pp. 396–403.
3. Cenziper (2015). For an extensive account, see also 'Abusive Head Trauma', Wikipedia. https://en.wikipedia.org/wiki/Abusive_head_trauma (accessed 7 September 2017).
4. https://www.judiciary.gov.uk/wp-content/uploads/2016/11/squier-v-gmc-protected-approved-judgment-20160311–2.pdf (accessed 27 October 2017).
5. 'Doctor Misled Courts in "Shaken Baby" Cases', BBC News, 11 March 2016. http://www.bbc.co.uk/news/uk-england-oxfordshire-35787095 (accessed 7 September 2017).
6. 'General Medical Council Behaving Like a Modern Inquisition', *Guardian*, 21 March 2016. https://www.theguardian.com/society/2016/mar/21/gmc-behaving-like-a-modern-inquisition-by-striking-off-dr-waney-squier (accessed 27 October 2017).
7. 'Should Waney Squier Have Been Struck Off Over Shaken Baby Syndrome?', *Newsnight*, 17 October 2016. http://www.bbc.co.uk/news/health–37672451 (accessed 7 September 2017).
8. Seymour J. Gray, John A. Benson Jr, Robert W. Reifenstein and Howard M. Spiro, 'Chronic Stress and Peptic Ulcer', *Journal of the American Medical Association* 147(16), pp. 1529–37.
9. Self-experimentation and self-prescription are themselves the subject of controversy. Some scientific ethics committees have reservations about self-experimentation; morally, however, it is significantly more acceptable than inflicting diseases on others. See also Esther Landhuis (2016), 'Do It Yourself? When the Researcher Becomes the Subject', *Science*, 5 December. http://www.sciencemag.org/careers/2016/12/do-it-yourself-when-researcher-becomes-subject (accessed 7 September 2017).
10. 'Office of the Nobel Laureates in Western Australia'. https://www.helicobacter.com/ (accessed 7 September 2017).
11. Pentti Sipponen and Barry J. Marshall (2000), 'Gastritis and Gastric Cancer: Western Countries', *Gastroenterology Clinics* 29(3), pp. 579–92. https://www.helicobacter.com/single-post/2000/01/01/Gastritis-and-Gastric-Cancer---Western-Countries (accessed 7 September 2017).
12. Tom Wildie (2017), 'Latest Helicobacter Pylori Breakthrough Could Lead to Eradication of Bacteria', ABC News, 4 April. http://www.abc.net.au/news/2017–04–04/researchers-who-discovered-heliobacter-learn-more-about-bacteria/8415686 (accessed 7 September 2017).
13. David Wootton (2013), *Galileo: Watcher of the Skies*, New Haven and London: Yale University Press.

14. For an amusing account, see Lydia Kang and Nate Pedersen, *Quackery: A Brief History of the Worst Ways to Cure Everything*, New York: Workman Publishing.
15. See also Sushil Bikhchandani, Amitabh Chandra, Dana Goldman and Ivo Welch (2002), 'The Economics of Iatroepidemics and Quackeries: Physician Learning, Informational Cascades and Geographic Variation in Medical Practice', Hanover, NH: Department of Economics, Dartmouth College.
16. What Michael Gove actually said is recorded on a YouTube video uploaded by rpmackay in 2016: 'Gove: Britons "Have Had Enough of Experts"'. https://www.youtube.com/watch?v=GGgiGtJk7MA (accessed 7 September 2017).
17. Michael Deacon (2016), 'Michael Gove's Guide to Britain's Greatest Enemy . . . the Experts', *Daily Telegraph*, 10 June. http://www.telegraph.co.uk/news/2016/06/10/michael-goves-guide-to-britains-greatest-enemy-the-experts/ (accessed 27 October 2017).
18. The origins and consequences of our disenchantment with experts, and the unfortunate concatenation of populism and anti-intellectualism, is explored by Tom Nichols (2017) in the aptly (if worryingly) titled book, *The Death of Expertise: The Campaign Against Established Knowledge and Why It Matters*, Oxford University Press.
19. For the BBC Trust report see 'Trust Conclusions on the Executive Report on Science Impartiality Review Actions', BBC Trust, July 2014. http://downloads.bbc.co.uk/bbctrust/assets/files/pdf/our_work/science_impartiality/trust_conclusions.pdf (accessed 7 September 2017). See also Emily Atkin (2014), 'To Improve Accuracy, BBC Tells Its Reporters to Stop Giving Air Time to Climate Deniers', *ThinkProgress*, 7 July. https://thinkprogress.org/to-improve-accuracy-bbc-tells-its-reporters-to-stop-giving-air-time-to-climate-deniers-c4b50fa1dddf/ (accessed 7 September 2017).
20. Tom Nichols emphasises the dangers of this in terms of the ongoing progress of scientific research. Nichols (2017).
21. For a general assessment, see Ben Goldacre (2009), *Bad Science*, London: HarperCollins.
22. For a survey of this literature, see Surowiecki (2004).
23. See Beryl Lieff Benderly (2016), 'How Scientific Culture Discourages New Ideas', *Science*, 6 July. http://www.sciencemag.org/careers/2016/07/how-scientific-culture-discourages-new-ideas (accessed 7 September 2017).
24. George Akerlof (1970), 'The Market for Lemons: Quality Uncertainty and the Market Mechanism', *Quarterly Journal of Economics* 84(3), pp. 488–500.
25. Brian Deer (2011), 'How the Case Against the MMR Vaccine was Fixed', *British Medical Journal* 342, pp. 77–82; Fiona Godlee, Jane Smith and Harvey Marcovitch (2011), 'The Fraud Behind the MMR Scare', *British Medical Journal* 342, pp. 64–6.
26. Matthias R. Effinger and Mattias K. Polborn (2001), 'Herding and Anti-Herding: A Model of Reputational Differentiation', *European Economic Review* 45(3), pp. 385–403.
27. A good summary of the key elements is in Yudhijit Bhattacharjee (2013), 'The Mind of a Con Man', *New York Times*, 26 April. https://www.

nytimes.com/2013/04/28/magazine/diederik-stapels-audacious-academic-fraud.html?pagewanted=all (accessed 29 March 2018).
28. Gustavo Saposnik, Jorge Maurino, Angel P. Sempere, Christian C. Ruff and Philippe N. Tobler (2017), 'Herding: A New Phenomenon Affecting Medical Decision-Making in Multiple Sclerosis Care? Lessons Learned from DIScUTIR MS', *Patient Preference and Adherence* 11, pp. 175–80.
29. Alan D. Sokal (1996), 'A Physicist Experiments with Cultural Studies', *Lingua Franca*. http://www.physics.nyu.edu/sokal/lingua_franca_v4/lingua_franca_v4.html (accessed 5 September 2017). For the article, see Alan D. Sokal (1994), 'Transgressing the Boundaries: Towards a Transformative Hermeneutics of Quantum Gravity', *Social Text* 46/47, pp. 217–52. For the journal editors' response to the hoax, see Bruce Robbins and Andrew Ross (1994), 'Editorial Response to Alan Sokal's Claim . . .', *Social Text*. http://www.physics.nyu.edu/sokal/SocialText_reply_LF.pdf (accessed 7 September 2017). See also 'Sokal affair', Wikipedia. https://en.wikipedia.org/wiki/Sokal_affair (accessed 7 September 2017).
30. Tversky and Kahneman (1974).
31. Michael Weisberg (2013), 'Modeling Herding Behavior and Its Risks', *Journal of Economic Methodology* 20(1), pp. 6–18; Ryan Muldoon and Michael Weisberg (2011), 'Epistemic Landscapes and the Division of Cognitive Labor', *Philosophy of Science* 76(2), pp. 225–52.
32. Thomas S. Kuhn (1996), *The Structure of Scientific Revolutions*, 3rd edn, University of Chicago Press.
33. Studying real juries is precluded for legal reasons.
34. Michelle Baddeley and Sophia Parkinson (2012), 'Group Decision-Making: An Economic Analysis of Social Influence and Individual Difference in Experimental Juries', *Journal of Socioeconomics* 41(5), pp. 558–73.
35. Beauty parades are a different phenomenon to the beauty contests which we explored in the last chapter in the context of financial speculation. In a beauty parade, a series of business people present their business plans to a committee – a board of directors for example. In the case of this FCA study the presentations would be to an oversight committee.
36. Tilba, Baddeley and Liao (2016).
37. Elinor Ostrom (2015), *Governing the Commons: The Evolution of Institutions for Collective Action*, Cambridge University Press.

8 Following the leader

1. 'Pied Piper of Hamelin', Wikipedia. https://en.wikipedia.org/wiki/Pied_Piper_of_Hamelin (accessed 7 September 2017); 'The Disturbing True Story of the Pied Piper of Hamelin', *Ancient Origins*, 14 August 2014. http://www.ancient-origins.net/myths-legends/disturbing-true-story-pied-piper-hamelin-001969?nopaging=1 (accessed 7 September 2017).
2. Peter Bazalgette explores the darker side of our identities in the context of genocides and mass murder, focusing on some of the atrocities of the twentieth century, including the Holocaust: Bazalgette (2017).
3. This contrasts with modern views from neuroscientists, for example Ramsey Raafat and colleagues who, in their analysis of human herding, argued that the leader is not essential to the herding phenomenon. See Raafat, Chater and Frith (2009).

4. Sigmund Freud (1921), p. 121.
5. Schumpeter (1934/1981).
6. Heinrich von Stackelberg (1934/2011), *Market Structure and Equilibrium* [*Marktform und Gleichgewicht*], trans. Damien Bazin, Rowland Hill and Lynn Urch, New York: Springer-Verlag.
7. Harold Hotelling (1929), 'Stability in Competition', *Economic Journal* 39(153), pp. 41–57.
8. Andrew Beer and Terry Clower (2014), 'Mobilizing Leadership in Cities and Regions', *Regional Studies, Regional Science* 1(1), pp. 5–20.
9. Michael Nye and Tom Hargreaves (2010), 'Exploring the Social Dynamics of Pro-Environmental Behaviour Change: A Comparative Study of Intervention Processes at Home and Work', *Journal of Industrial Ecology* 14(1), pp. 137–49.
10. Andrea Galeotti and Sanjeev Goyal (2010), 'The Law of the Few', *American Economic Review* 100(4), pp. 1468–92.
11. For an engaging exploration of these influences in the business and commercial world, see Julia Hobsbawm (2017), *Fully Connected: Surviving and Thriving in the Age of Overload*, London: Bloomsbury. For an analysis from the perspective of economic theory, see Sanjeev Goyal (2010), *Connections: An Introduction to the Economics of Networks*, Princeton University Press.
12. Sarah Harris (2017), 'Under the Influence', *Vogue*, March, pp. 318–23.
13. Milgram (1963); Adam Cohen (2008), 'Four Decades After Milgram, We're Still Willing to Inflict Pain', *New York Times*, 28 December. http://www.nytimes.com/2008/12/29/opinion/29mon3.html?mcubz=1 (accessed 7 September 2017). Recent analysis of Milgram's early experiments has suggested that the interpretation of the findings may have been flawed: he may have attributed too much to his hypothesis of obedience to authority. See, for example, Adam Sherwin (2014), 'Famous Milgram "Electric Shocks" Experiment Drew Wrong Conclusions About Evil, Say Psychologists', *Independent*, 4 September. http://www.independent.co.uk/news/science/famous-milgram-electric-shocks-experiment-drew-wrong-conclusions-about-evil-say-psychologists-9712600.html (accessed 7 September 2017).
14. Saul McLeod (2007), 'The Milgram Experiment', *Simply Psychology*. http://www.simplypsychology.org/milgram.html (accessed 7 September 2017).
15. Milgram (1963). See also Stanley Milgram (1974), *Obedience to Authority*, New York: Harper and Row.
16. Marcus Cheetham, Andreas F. Pedroni, Angus Antley, Mel Slater and Lutz Jäncke (2009), 'Virtual Milgram: Empathic Concern or Personal Distress? Evidence from Functional MRI and Dispositional Measures', *Frontiers in Human Neuroscience* 3, article ID 29. https://doi.org/10.3389/neuro.09.029.2009 (accessed 7 September 2017). See also Mel Slater, Angus Antley, Adam Davison, David Swapp, Christoph Guger, Chris Barker, Nancy Pistrang and Maria V. Sanchez-Vives (2006), 'A Virtual Reprise of the Stanley Milgram Obedience Experiments,' *PLOS ONE* 1(1):e39. https://doi.org/10.1371/journal.pone.0000039 (accessed 7 September 2017).
17. Philip Zimbardo (2008), *The Lucifer Effect: How Good People Turn Evil*, London: Rider/Random House. See also Craig Haney, Curtis Banks

and Philip Zimbardo (1973), 'A Study of Prisoners and Guards in a Simulated Prison', *Naval Research Review* 30, pp. 4–17. http://www.garysturt.free-online.co.uk/zimbardo.htm (accessed 7 September 2017); and Philip Zimbardo (1999–2017), 'Frequently Asked Questions', *Stanford Prison Experiment*. http://www.prisonexp.org/faq (accessed 7 September 2017).

18. Thomas Carnahan and Sam McFarland (2007), 'Revisiting the Stanford Prison Experiment: Could Participant Self-Selection Have Led the Cruelty?', *Personality and Social Psychology Bulletin* 33(5), pp. 603–14.
19. Craig Haney, Curtis W. Banks and Philip G. Zimbardo (1973), 'Interpersonal Dynamics in a Simulated Prison', *International Journal of Criminology and Penology* 1, pp. 69–97.
20. Zimbardo (2008).
21. Armen A. Alchian and Harold Demsetz (1972), 'Production, Information Costs, and Economic Organization', *American Economic Review* 62(5), pp. 777–95.
22. Nye and Hargreaves (2010).
23. Danny Wallace (2004), *Join Me: The True Story of a Man Who Started a Cult by Accident*, London: Ebury Press.
24. Sigmund Freud (1930), *Civilization and Its Discontents*, in *The Standard Edition of the Complete Psychological Works of Sigmund Freud*, Vol. XXI (1927–31), trans. James Strachey in collaboration with Anna Freud, London: Vintage, p. 64.
25. Sam Harris, Jonas T. Kaplan, Ashley Curiel, Susan Y. Bookheimer, Marco Iacoboni and Mark S. Cohen (2009), 'The Neural Correlates of Religious and Nonreligious Belief', *PLoS ONE* 4(10): e7272. https://doi.org/10.1371/journal.pone.0007272 (accessed 27 October 2017).
26. Le Bon (1895), p. 12.
27. Stephen Quirke (2014), *Exploring Religion in Ancient Egypt*, London: John Wiley and Sons. See also Alastair Sook (2014), 'Akhenaten – Mad, Bad or Brilliant?', *Daily Telegraph*, 9 January. http://www.telegraph.co.uk/culture/art/10561090/Akhenaten-mad-bad-or-brilliant.html (accessed 7 September 2017).
28. The use of the phrase 'sun-disk' captures the idea that the sun has a face, i.e. a disk, so the sun-disk is the face of the sun. For some more on bizarre cult teachings, see Debra Kelly (2013), '10 Bizarre Cult Teachings', *Listverse*, 4 December. http://listverse.com/2013/12/04/10-bizarre-cult-teachings (accessed 7 September 2017).
29. Janet Reitman (2013), *Inside Scientology: The Story of America's Most Secretive Religion*, Boston, MA: Mariner Books/Houghton Mifflin Harcourt.
30. Lou Manza (2016), 'How Cults Exploit One of Our Most Basic Psychological Urges', *The Conversation*, 14 April. https://theconversation.com/how-cults-exploit-one-of-our-most-basic-psychological-urges–57101 (accessed 27 October 2017); Mark Banschick (2013), 'What Awful Marriages and Cults Have in Common', *Psychology Today*, 28 May. https://www.psychologytoday.com/blog/the-intelligent-divorce/201305/what-awful-marriages-cults-have-in-common (accessed 27 October 2017).
31. Dorian Lynskey (2013), 'Beatlemania: "The Screamers" and Other Tales of Fandom', *Guardian*, 29 September. https://www.theguardian.com/

music/2013/sep/29/beatlemania-screamers-fandom-teenagers-hysteria (accessed 7 September 2017).
32. Martin Creasy (2010), *Beatlemania!: The Real Story of The Beatles UK Tours 1963–1965*, London: Omnibus Press. See also Joli Jensen (1992), 'Fandom as Pathology: The Consequences of Characterisation', in Lisa A. Lewis (ed.) (1992), *The Adoring Audience: Fan Culture and Popular Media*, London: Routledge, pp. 9–29.
33. O.G.T. Sonneck (1922), 'Henrick Heine's Musical Feuilletons', *Musical Quarterly* 8, pp. 457–8.
34. T.R.J. Nicholson, C. Pariante and D. McLoughlin (2009), 'Stendhal Syndrome: A Case of Cultural Overload', *British Medical Journal* case reports.
35. Keynes (1936), p. 374.
36. Bikhchandani, Hirshleifer and Welch (1998).
37. 'Extreme Tweeting', *The Economist*, 21 November 2015, p. 43.
38. Ambrose Evans-Pritchard (2016), '"Irritation and Anger" May Lead to Brexit, Says Influential Psychologist', *Daily Telegraph*, 6 June. http://www.telegraph.co.uk/business/2016/06/05/british-voters-succumbing-to-impulse-irritation-and-anger---and/ (accessed 14 June 2017).
39. Hunt Allcott and Matthew Gentzkow (2017), 'Social Media and Fake News in the 2016 Election', *Journal of Economic Perspectives* 31(2), pp. 211–36.
40. 'Trump's Executive Order: Who Does Travel Ban Affect?', BBC News, 10 February 2017. http://www.bbc.com/news/world-us-canada-38781302 (accessed 10 February 2017).
41. 'Donald Trump's File', PolitiFact. http://www.politifact.com/personalities/donald-trump/ (accessed 5 September 2017).
42. 'Hillary Clinton's File', PolitiFact. http://www.politifact.com/personalities/hillary-clinton/ (accessed 7 September 2017).
43. For an assessment of social media, see 'Social Media's Threat to Democracy', *The Economist*, 4 November 2017.
44. Sunstein (2002).
45. Craig Silverman (2016), 'This Analysis Shows How Viral Fake Election News Stories Outperformed Real News on Facebook', BuzzFeed News, 17 November. https://www.buzzfeed.com/craigsilverman/viral-fake-election-news-outperformed-real-news-on-facebook?utm_term=.fqNdd13EE#.ljw44EBpp; Timothy B. Lee (2016), 'The Top 20 Fake News Stories Outperformed Real News at the End of the 2016 Campaign', *Vox*, 16 November. https://www.vox.com/new-money/2016/11/16/13659840/facebook-fake-news-chart (accessed 18 November 2016).
46. Cass R. Sunstein (2017), *#Republic: Divided Democracy in the Age of Social Media*, Princeton University Press.
47. For example, see Seth Flaxman, Sharad Goeal and Justin M. Rao (2016), 'Filter Bubbles, Echo Chambers, and Online News Consumption', *Public Opinion Quarterly* 80, pp. 298–320.
48. See also Fil Menzer (2016), 'The Spread of Misinformation in Social Media', Northwestern Institute on Complex Systems seminar. http://cnets.indiana.edu/blog/2016/10/10/spread-of-misinformation/ (accessed 7 September 2017).

Conclusion: Copycats versus contrarians

1. Social capital is a concept first explored by sociologists and psychologists: for example, see Jane Jacobs (1961), *The Death and Life of Great American Cities*, New York: Random House; James S. Coleman (1988), 'Social Capital in the Creation of Human Capital', *American Journal of Sociology* 94, pp. 95–120; Robert D. Putnam (1993), 'What Makes Democracy Work?', *National Civic Review* 82(2), pp. 101–7; and David Halpern (2005), *Social Capital*, Cambridge: Polity Press. For a survey of some of the economists' insights about social capital, particularly in the context of developing economies, see Partha Dasgupta and Ismail Serageldin (eds) (2010), *Social Capital: A Multifaceted Perspective*, Washington, DC: The International Bank for Reconstruction and Development/World Bank. Some of this literature is summarised for a general reader in Partha Dasgupta (2007), *Economics: A Very Short Introduction*, Oxford University Press.
2. Mark S. Granovetter (1973), 'The Strength of Weak Ties', *American Journal of Sociology* 78(6), pp. 1360–80.
3. Robert Putnam introduced a parallel between strong ties and weak ties as bonding versus bridging forms of social capital. Strong ties bond people closely; weak ties are more like bridges that connect people more gently. See Robert D. Putnam (2000), *Bowling Alone: The Collapse and Revival of American Community*, New York/London: Simon and Schuster.
4. Thaler and Sunstein (2008), in particular ch. 3, 'Following the herd'.
5. For an analysis of nudges from the perspective of psychologist David Halpern, director of the UK's Behavioural Insights Team, a.k.a. 'the Nudge Unit', see David Halpern (2015), *Inside the Nudge Unit: How Small Changes Can Make a Big Difference*, London: Ebury Press.
6. Sunstein (2002).

Further reading

Introduction

Earls, Mark (2009). *Herd: How to Change Mass Behavior by Harnessing Our True Nature*, Chichester: John Wiley and Sons.
Surowiecki, James (2004). *The Wisdom of Crowds: Why the Many Are Smarter Than the Few*, London: Abacus Books.

1 Clever copying

Akerlof, George A. and Rachel E. Kranton (2000). 'Economics and Identity', *Quarterly Journal of Economics* 115(3), pp. 715–53.
— (2011). *Identity Economics: How Our Identities Shape Our Work, Wages, and Well-Being*, Princeton University Press.
Anderson, Lisa R. and Charles A. Holt (1996). 'Classroom Games: Information Cascades', *Journal of Economic Perspectives* 10(4), pp. 187–93.
— (1997). 'Information Cascades in the Laboratory', *American Economic Review* 87(5), pp. 847–62.
Banerjee, Abhijit (1992). 'A Simple Model of Herd Behavior', *Quarterly Journal of Economics* 107(3), pp. 797–817.
Bernheim, B. Douglas (1994). 'A Theory of Conformity', *Journal of Political Economy* 102(5), pp. 841–77.
Bikhchandani, Sushil, David Hirshleifer and Ivo Welch (1992). 'A Theory of Fads, Fashion, Custom, and Cultural Change as Informational Cascades', *Journal of Political Economy* 100(5), pp. 992–1026.
— (1998). 'Learning from the Behavior of Others: Conformity, Fads, and Informational Cascades', *Journal of Economic Perspectives* 12(3), pp. 151–70.
Chamley, Christophe P. (2003). *Rational Herds: Economic Models of Social Learning*, Cambridge University Press.
Leibenstein, Harvey (1950). 'Bandwagon, Snob, and Veblen Effects in the Theory of Consumers' Demand', *Quarterly Journal of Economics* 64(2), pp. 183–207.

2 Mob psychology

Asch, Solomon (1952). *Social Psychology*, Englewood Cliff, NJ: Prentice-Hall.
— (1955), 'Opinions and Social Pressure', *Scientific American* 193(5), pp. 31–5.
Bond, Rod and Peter B. Smith (1996). 'Culture and Conformity: A Meta-Analysis of the Studies Using Asch's (1952b, 1956) Line Judgment Task', *Psychological Bulletin* 119(1), pp. 111–37.
Cialdini, Robert B. (2007). *Influence: The Psychology of Persuasion*, New York: HarperCollins.
Freud, Sigmund (1991). *Civilization, Society and Religions: 'Group Psychology and the Analysis of the Ego', 'Future of An Illusion' and 'Civilization and Its Discontents'*, Penguin Freud Library, London: Penguin Books.
— (2010). *Sigmund Freud Collected Works: The Psychopathology of Everyday Life, The Theory of Sexuality, Beyond the Pleasure Principle, The Ego and the Id, and The Future of an Illusion*, trans. A.A. Brill, Seattle, WA: Pacific Publishing Studio.
Le Bon, Gustave (1895). *The Crowd: A Study of the Popular Mind*, Lexington, KY: Maestro Reprints.
Mackay, Charles (1841). *Extraordinary Popular Delusions and the Madness of Crowds*, Lexington, KY: Maestro Reprints.
Reich, Wilhelm (1946). *The Mass Psychology of Fascism*, trans. Theodore P. Wolfe, New York: Orgone Institute Press.
Shiller, Robert J. (1995). 'Conversation, Information and Herd Behavior', *American Economic Review* 85(2), pp. 181–5.
Tajfel, Henri (1970). 'Experiments in Intergroup Discrimination', *Scientific American* 223, pp. 96–102.
Tajfel, Henri, M.G. Billig, R.P. Bundy and Claude Flament (1971). 'Social Categorization and Intergroup Behaviour', *European Journal of Social Psychology* 1(2), pp. 149–78.

3 Herding on the brain

Baddeley, Michelle (2010). 'Herding, Social Influence and Economic Decision-Making: Socio-Psychological and Neuroscientific Analyses', *Philosophical Transactions of the Royal Society B* 365(1538), pp. 281–90.
Burke, Christopher, Michelle Baddeley, Philippe Tobler and Wolfram Schultz (2010). 'Striatal BOLD Response Reflects the Impact of Herd Information on Financial Decisions', *Frontiers in Human Neuroscience* 4, article 48. https://doi.org/10.3389/fnhum.2010.00048 (accessed 5 September 2017).
Camerer, Colin F., George Loewenstein and Drazen Prelec (2005). 'Neuroeconomics: How Neuroscience Can Inform Economics', *Journal of Economic Literature* 43(1), pp. 9–64.
Damasio, Antonio (1994/2006). *Descartes' Error: Emotion, Reason and the Human Brain*, London: Vintage.
Hurley, Susan and Nick Chater (eds) (2005). *Perspectives on Imitation: From Neuroscience to Social Science*, Cambridge, MA: MIT Press.
Iacoboni, Marco, Roger P. Woods, Marcel Brass, Harold Bekkering, John C. Mazziotta and Giacomo Rizzolatti (1999). 'Cortical Mechanisms of Human Imitation', *Science* 286, pp. 2526–8.

Kahneman, Daniel (2011). *Thinking, Fast and Slow*, New York: Farrar, Straus and Giroux.
Klucharev, Vasily, Kaisa Hytönen, Mark Rijpkema, Ale Smidts and Guillén Fernández (2009). 'Reinforcement Learning Signal Predicts Social Conformity', *Neuron* 61(1), pp. 140–51.

4 Animal herds

Axelrod, Robert (1984). *The Evolution of Cooperation*, Cambridge, MA: Basic Books.
Blackmore, Susan (1999). *The Meme Machine*, Oxford University Press.
Cohen, Jonathan D. (2005). 'The Vulcanization of the Human Brain: A Neural Perspective on Interactions between Cognition and Emotion', *Journal of Economic Perspectives* 19(4), pp. 3–24.
Danchin, Étienne, L.-S. Giraldeau, T.J. Valone and R.H. Wagner (2004). 'Public Information: From Nosy Neighbours to Cultural Evolution', *Science* 305, pp. 487–91.
Darwin, Charles (1859/2011). *On the Origin of Species* (Collins Classics edn), New York: HarperCollins.
Davies, Nicholas B., John R. Krebs and Stuart A. West (2012). *An Introduction to Behavioural Ecology* (4th edn), Oxford: Wiley Blackwell.
Dawkins, Richard (1976). *The Selfish Gene*, Oxford University Press.
Gould, Stephen Jay (2001). *The Lying Stones of Marrakech: Penultimate Reflections in Natural History*, London: Vintage.
Kirman, Alan (1993). 'Ants, Rationality and Recruitment', *Quarterly Journal of Economics* 108(1), pp. 137–56.
Lynch, Aaron (1996). *Thought Contagion: How Belief Spreads Through Society*, New York: Basic Books.
Maynard Smith, John (1974). 'The Theory of Games and the Evolution of Animal Conflicts', *Journal of Theoretical Biology* 47, pp. 209–21.
Morris, Desmond (1967). *The Naked Ape: A Zoologist's Study of the Human Animal*, London: Jonathan Cape.
Raafat, Ramsey M., Nick Chater and Chris Frith (2009). 'Herding in Humans', *Trends in Cognitive Sciences* 13(10), pp. 420–8.
Rizzolatti, Giacomo and Laila Craighero (2004). 'The Mirror Neuron System', *Annual Review of Neuroscience* 27(1), pp. 169–92.
Safina, Carl (2015). *Beyond Words: What Animals Think and Feel*, New York: John Macrae/Henry Holt and Company.
Shermer, Michael (2009). *The Mind of the Market: How Biology and Psychology Shape Our Economic Lives*, New York: Henry Holt and Company.
Simon, Herbert (1990). 'A Mechanism for Social Selection and Successful Altruism', *Science* 250, pp. 1665–8.
Whitehead, Hal and Luke Rendell (2015). *The Cultural Lives of Whales and Dolphins*, University of Chicago Press.
Wilson, Edward O. (1975). *Sociobiology: The New Synthesis*, Cambridge, MA: Harvard University Press.
Wilson, Edward O. and Bert Hölldobler (2005). 'Eusociality: Origin and Consequences', *Proceedings of the National Academy of Sciences* 102(38), pp. 13367–71.

5 Mavericks

Bernheim, B. Douglas (1994). 'A Theory of Conformity', *Journal of Political Economy* 102(5), pp. 841–77.
Galeotti, Andrea and Sanjeev Goyal (2010). 'The Law of the Few', *American Economic Review* 100(4), pp. 1468–92.
Hirshleifer, David and Robert Noah (1998). 'Misfits and Social Progress', in Robert Noah (1998), *Essays in Learning and the Revelation of Private Information*, PhD Thesis, University of Michigan: ProQuest Dissertations Publishing.
Sunstein, Cass R. (2002). 'Conformity and Dissent', John M. Olin Law and Economics Working Paper No. 164, The Law School, University of Chicago. http://chicagounbound.uchicago.edu/public_law_and_legal_theory/68/ (accessed 5 September 2017).
Weightman, Gavin (2015). *Eureka: How Invention Happens*, New Haven and London: Yale University Press.

6 Entrepreneurs versus speculators

Acemoğlu, Daron (1992). 'Learning about Others' Actions and the Investment Accelerator', *Economic Journal* 103(417), pp. 318–28.
Akerlof, George and Robert Shiller (2009). *Animal Spirits: How Human Psychology Drives the Economy and Why It Matters for Global Capitalism*, Princeton University Press.
Avery, Christopher and Peter Zemsky (1998). 'Multidimensional Uncertainty and Herd Behavior in Financial Markets', *American Economic Review* 88(4), pp. 724–48.
Baddeley, Michelle (2018). 'Financial Instability and Speculative Bubbles: Behavioural Insights and Policy Implications', in *Alternative Approaches in Macroeconomics: Essays in Honour of John McCombie*, ed. Philip Arestis, London: Palgrave Macmillan, pp. 209–34.
Butler, Timothy (2017). 'Hiring an Entrepreneurial Leader', *Harvard Business Review*, March–April, pp. 84–93,
Chancellor, Edward (1998). *Devil Take the Hindmost: A History of Financial Speculation*, New York/London: Plume/Penguin Books.
Devenow, Andrea and Ivo Welch (1996). 'Rational Herding in Financial Economics', *European Economic Review* 40, pp. 603–15.
Drehmann, Mathias, Jörg Oechssler and Andreas Roider (2005). 'Herding and Contrarian Behavior in Financial Markets: An Internet Experiment', *American Economic Review* 95(5), pp. 1403–26.
Garber, Peter M. (2001). *Famous First Bubbles: The Fundamentals of Early Manias*, Cambridge, MA: MIT Press.
Ingham, Geoffrey (2013). *The Nature of Money*, Cambridge: Polity Press.
Keynes, John Maynard (1936). *The General Theory of Employment, Interest and Money*, London: Macmillan and the Royal Economic Society; see especially chapter 12.
— (1937). 'The General Theory of Employment', *Quarterly Journal of Economics* 51(2), pp. 209–23.
Kindleberger, Charles P. and Robin Aliber (2005). *Manias, Panics and Crashes: A History of Financial Crises* (5th edn), Hoboken, NJ: John Wiley and Sons.

Landes, David S., Joel Mokyr and William J. Baumol (eds) (2012). *The Invention of Enterprise: Entrepreneurship from Ancient Mesopotamia to Modern Times*, Princeton University Press.
Lo, Andrew W., Dmitry V. Repin and Brett N. Steenbarger (2005). 'Fear and Greed in Financial Markets: A Clinical Study of Day Traders', *American Economic Review* 95(2), pp. 352–9.
Loewenstein, George, Elke U. Weber, Christopher K. Hsee and Ned Welch (2001). 'Risk as Feelings', *Psychological Bulletin* 127(2), pp. 267–86.
Mackay, Charles (1841). *Extraordinary Popular Delusions and the Madness of Crowds*, Lexington, KY: Maestro Reprints.
Prechter, Robert (2016). *The Socionomic Theory of Finance*, Gainesville, GA: Socionomics Institute Press.
Scharfstein, D.S. and J.C. Stein (1990). 'Herd Behavior and Investment', *American Economic Review* 80(2), pp. 465–79.
Topol, Richard (1991). 'Bubbles and Volatility of Stock Prices: Effect of Mimetic Contagion', *Economic Journal* 101(407), pp. 786–800.

7 Herding experts

Baddeley, Michelle (2013). 'Herding, Social Influence and Expert Opinion', *Journal of Economic Methodology* 20, pp. 37–45.
— (2015). 'Herding, Social Influences and Behavioural Bias in Scientific Research', *European Molecular Biology Organisation Reports* 16(8), pp. 902–5.
— (2017). 'Experts in Policy Land: Insights from Behavioral Economics on Improving Experts' Advice for Policy-Makers', *Journal of Behavioral Economics for Policy* 1(1), pp. 27–31.
Baddeley, Michelle, Andrew Curtis and Rachel Wood (2004). 'An Introduction to Prior Information Derived from Probabilistic Judgments; Elicitation of Knowledge, Cognitive Bias and Herding', in *Geological Prior Information: Informing Science and Engineering*, ed. A. Curtis and R. Wood, Geological Society, London, Special Publications 239, pp. 15–27.
Deer, Brian (2011). 'How the Case Against the MMR Vaccine was Fixed', *British Medical Journal* 342, pp. 77–82.
Kuhn, Thomas S. (1996). *The Structure of Scientific Revolutions* (3rd edn), University of Chicago Press.
Nichols, Tom (2017). *The Death of Expertise: The Campaign Against Established Knowledge and Why It Matters*, Oxford University Press.
Surowiecki, James (2004). *The Wisdom of Crowds: Why the Many Are Smarter Than the Few*, London: Abacus.
Weisberg, Michael (2013). 'Modeling Herding Behavior and Its Risks', *Journal of Economic Methodology* 20(1), pp. 6–18.

8 Following the leader

Alchian, Armen A. and Harold Demsetz (1972). 'Production, Information Costs, and Economic Organization', *American Economic Review* 62(5), pp. 777–95.
Bazalgette, Peter (2017). *The Empathy Instinct: How to Create a More Civil Society*, London: John Murray.
Milgram, Stanley (1963). 'Behavioral Study of Obedience', *Journal of Abnormal and Social Psychology* 67, pp. 371–8.

— (1974). *Obedience to Authority*, New York: Harper and Row.
Wallace, Danny (2004). *Join Me: The True Story of a Man Who Started a Cult by Accident*, London: Ebury Press.
Zimbardo, Philip (2008). *The Lucifer Effect: How Good People Turn Evil*, London: Rider/Random House.

Conclusion: Copycats versus contrarians

Granovetter, Mark S. (1973). 'The Strength of Weak Ties', *American Journal of Sociology* 78(6), pp. 1360–80.
Halpern, David (2015). *Inside the Nudge Unit: How Small Changes Can Make a Big Difference*, London: Ebury Press.
Harari, Yuval Noah (2014). *Sapiens: A Brief History of Humankind*, London: Harvill Secker/Random House.
Maynard Smith, John (1982). *Evolution and the Theory of Games*, Cambridge University Press.
Morris, Desmond (1969/1994). *The Human Zoo*, London: Vintage.
Seabright, Paul (2004). *The Company of Strangers: A Natural History of Economic Life*, Princeton University Press.
Thaler, Richard and Cass Sunstein (2008). *Nudge: Improving Decisions about Health, Wealth, and Happiness*, New Haven and London: Yale University Press.

Acknowledgements

I am grateful to many people, especially to Taiba Batool and her team at Yale University Press in London. A few years back I sketched out some ideas that now form the bones of this book. Subsequently and serendipitously, Taiba approached me about writing a book on very similar themes. She has expertly guided this book through all its stages, giving me plenty of encouragement, advice and ideas along the way. I am also grateful to my copy-editors, Rachael Lonsdale and Jacob Blandy. The manuscript that first landed on their desk was much flawed. Their efforts, perseverance and attention to detail helped me to transform it into something immeasurably better. Any remaining errors, omissions and infelicities are mine. I would also like to thank various colleagues and Yale's anonymous reviewers for their helpful and constructive comments, which helped me to focus more clearly on what I am really trying to say. My thanks to my co-authors Christopher Burke, Philippe Tobler and Wolfram Schultz for kindly agreeing to the reprinting of the images used in chapter 3, and for collaborating with me in this fascinating research. My gratitude also goes to the Leverhulme Trust for their financial sponsorship of our Neuroeconomics of

Herding research project, during which many of the ideas explored in this book were first formulated. I should also note my appreciation of organisations that provide free and advertising-free access to news and information, primarily Wikipedia but also BBC News Online and Guardian Online.

My enduring gratitude goes to my father and his enviable intellect and memory. He read the manuscript meticulously, was enormously enthusiastic about what I've written and gave me loads of invaluable tips and advice – especially helpful in writing the technical discussions of brain scanning, a field in which he was a pioneering radiologist. Perhaps, next time, we shall write a book together. His advice on books to read, and insights relating to neuroscience and evolutionary biology, were invaluable. His own books, some written under his pseudonym John Bates, are a fantastic resource as unique and wide-ranging accounts of the medical and physical sciences, as well as theology and intellectual history.

To my mother, as well as my father, I will always be grateful for encouragement and inspiration, for not seeming to worry about giving me a big head, and for engendering within me a spirit of curiosity. Last but not least, my eternal thanks go to my husband Chris for being a sounding board – even while suffering at the coal face of what probably seemed to him like my interminable efforts to get this book written. Few work-aholic wives can be as lucky as I am in enjoying their husband's unwavering support, patience and good humour.

Illustration credits

1 Andy Last. 4 Burke, Baddeley, Tobler and Schultz (2010), 'Striatal BOLD response reflects the impact of herd information on financial decisions', *Frontiers in Human Neuroscience*, article 48, http://journal.frontiersin.org/article/10.3389/fnhum.2010.00048/full. 5 Dave Burns. 6 Heath Robinson Museum. 7 Frans Hals Museum, Haarlem, Netherlands. 8 Photograph by Jean-Pierre Dalbéra from Paris, France.

Index